House-Training Your VCR

A Help Manual for Humans

by

Dave Murray

Illustrated by Joe Congel

Grapevine Publications, Inc.
P.O. Box 2449
Corvallis, Oregon   97339-2449

Acknowledgments

Many products and services mentioned in this book are trademarks of their respective companies, and as such are identified by initial capitalization. Any omissions or inconsistencies are oversights to be corrected in future editions. "Time Machines" is a registered trademark of Sony Corporation. "Dolby" is a registered trademark of Dolby Laboratories. "JVC" is a registered trademark of Victor Company of Japan, Ltd. "RCA" is a registered trademark of Thomson Consumer Electronics. "VCR Plus+" is a registered trademark of Gemstar Development Corporation. "Macrovision" is a registered trademark of Macrovision Corp. "HBO" and "Cinemax" are registered trademarks of Home Box Office, Inc., a subsidiary of Time/Warner, Inc. "The Movie Channel" and "Showtime" are registered trademarks of Showtime/The Movie Channel, Inc. "The Disney Channel" is a registered trademark of the Walt Disney Company. "CNN" is a registered trademark of the Turner Broadcasting Company. "ESPN" is a registered trademark of the Entertainment and Sports Programming Network. "C-SPAN" is a registered trademark of C-SPAN Cable Satellite Public Affairs Network. "MTV" is a registered trademark of MTV Networks, Inc.

Printed in the United States of America
ISBN 0-931011-36-1

First Printing — October, 1991

"We'd Like to Thank the Academy..." Dept.

Actually, we don't know anyone at the Academy, but we *would* like to acknowledge the substantial contributions the following people made in bringing this book from a germ of an idea to completion:

We're very grateful to Jeff Riha, Marlon Legaspi, Klee Dugan, Joe Scripa, Mike Greenstein, Carol Jeschke, Dana Jeschke, Tracy Bunis, Jim McClenthan, Carl Cafarelli, Dan Murray, Harmony Murray, Kathy Murray, Judy Colozzi, Guy Colozzi, Paul Hoffman, Julie LaVolpe, Reynolds Baughman, Jr., Steve Heller, David Heller, Jerry Heller, Paul Clark, Dave D. Gibbs Domroe, Randy Myers, Gary Steinebach, Greg Horvath, Todd Bitter, Gian Agonito, Rosemary Agonito, Chris and Dan at Grapevine, and everyone else who offered their support and encouragement.

Dave's special thanks: to my mom, Pauline Savage and to my dad and "other mom," Fred and Luise Murray; to Steve and Dede Piesciuk for their extra efforts; to David "Arthur Dietrich" Barany for taking, with patience and grace, all 755 of my phone calls as he helped me decipher the "instruction manual" that came with the word processing program this book was written on, which will go unnamed because I might say some bad things and get sued; to Jim Bouton, for **Ball Four**; and to Dave Barry, the funniest man in the history of the known universe.

Joe's special thanks: to Joshua McIlvain, who always believed, and was always interested enough to ask about "Cartoon Joe's" world; to my wife Jackie, very loving thanks for hanging in there, even when maybe she shouldn't have and for putting up with perhaps the laziest person in the world; and to my daughter, Rita, who just by merely existing brings more joy and happiness to my life than I ever thought possible.

This book is dedicated to the crew of store 50 in Dewitt (1988-1991). It was fun while it lasted.

In memory of Louis A. Colozzi.

CONTENTS:

1. HOW TO USE THIS BOOK

Do you secretly wonder if all VCR manuals are written by committee—five chimpanzees with typewriters? When you read your VCR manual, does it feel like standing for an hour in the wrong line at the Department of Motor Vehicles?

Then this book is for you.

VCRs have been the bane of mankind for hundreds of years.* Indeed, certain historians theorize that Vincent Van Gogh, the famous painter, sliced off his ear because he couldn't figure out how to time-shift-record with his VCR. And here are today's revealing statistics:

- A recent poll of VCR owners showed that 112% of them cannot program their machines to record.

- The same owners listed these as the *primary uses* of their VCRs:

80%:	Movie Rental Device
14%:	Digital Clock
12%:	Night Light
6%:	Nice Warm Place For The Cat To Sleep

*It seems like that long, anyway.

- Furthermore, over 75% of these same owners said that, if forced to choose, they would willingly undergo root-canal work **without anesthetic**, rather than read the instruction manual that comes with any consumer electronic device. The rest of those surveyed refused even to speculate.

Shocking? Indeed. It's an enormous social problem. This book can and will help—if you let it—but *don't* treat it as an instruction manual. You've got an instruction manual already—the one that came with the machine—probably with phrases like:

> *"Please will you take care to push the record time shift function activation button now. Yes. Activating the time shift. Now with careful, you' ll be. Yes."*

Even subscribers to *Pocket Protectors Digest* don't get too far with stuff like that. Instruction manuals are often poorly written and far more complicated than necessary.* Well, that's where this book comes in. Beneath its flip, obnoxious exterior, its information is actually quite clear and concise. The whole idea here is to break down that wall of technical fog that separates regular people and their VCRs.

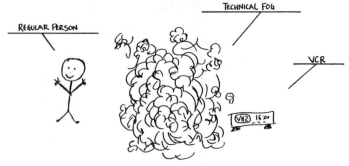

*For more extensive ranting and raving on this topic, please see pages 12-13.

The very best way to make **"House-Training Your VCR"** work for you —whether you own a VHS or a Beta machine—is by following the ground rules you see below:

THE GROUND RULES YOU SEE BELOW

1. **Use the Table of Contents or the Index:** Many people prefer instead to randomly hunt and search—like driving around without a road map—but sometimes that takes longer.

2. **Try Everything Once:** Put your VCR through its paces. After reading this book, you should never find yourself saying *"I don't know how to do that with mine—I've never tried it."* Which leads to...

3. **Don't Worry About Screwing Up**. You probably will anyway. But so what? You don't have to take out an ad in the paper about it: *"OWNER TOO STUPID TO USE VCR"*

    So, learn by doing. Remember: No matter how badly you mess up, you probably won't even ruin the carpet.

4. **Remember that VCRs are Just Stupid Machines.**

    Example: For some reason, you want to record "A Very Brady Christmas" at 9 o'clock on Friday. You won't be home then, but you find out that your friend, Dave, will be, so you ask him to do it. But you don't actually specify Friday *night*—no need to. You know that Dave, without a lot of research, can figure out that it must be on in the prime-time evening slot. But your VCR isn't as smart as Dave.* If you set its timer for **AM** instead of **PM**, you'll get Sally Jessy Raphael or something. But it isn't malfunctioning. It's just doing *exactly* what you tell it to do—no less, no more.

5. **Use the Glossary** (page 92): VCR manufacturers like to give identical features different names—to pretend their brand is the only one in the known universe with that marvelous function.

    Example: Most VCRs can be set to stop recording after a certain duration of time—like the sleep timer that turns off your clock radio after you drift off. But do VCR manufacturers simply call it a sleep timer? Of course not. They call it OTR, or XPR, or ITR, or some other easy-to-remember abbreviation.

*a tremendous source of pride for Dave.

To make matters worse, the language spoken by most VCR salespersons is a little-known dialect called **Digitenglish**.

DIGITENGLISH

Digitenglish is noted for its **H**eavy **A**cronym **R**eliance, or **HAR**. **HAR** greatly decreases the amount of space required for **CET** (**C**omplex **E**lectronic **T**erms), but it also tends to decrease understanding and increase incidence of **CRATTIBIP** (**C**onfused **R**eader **A**ngrily **T**earing **T**he **I**nstruction **B**ook **I**nto **P**ieces.)

SECOND-DEGREE CRATTIBIP

Resist the temptation! Although instruction manuals laced with Digitenglish do make terrific confetti, you should keep yours intact. To avoid accidental, involuntary **CRATTIBIP**ping, use the Glossary (page 92).

6. **<u>Use</u> your Instruction Manual**. Yes—it *is* a pain. You'd rather just muddle through, count on common sense to guide you well enough, and if that doesn't work, *then* you can always try the manual—if you haven't already **CRATTIBIP**ped it on principle.

Unfortunately, your manual is still more specific to your machine than this book can be. The purpose of this House-Training book is to *translate* your manual, really. Once you understand your VCR and its obscure terminology, you'll read your manual with a new sense of confidence—fun for the whole family. Here are some highlights to look for:

— It assumes you understand all of the Digitenglish, but you get **detailed instructions** on how to put the plug into the wall:

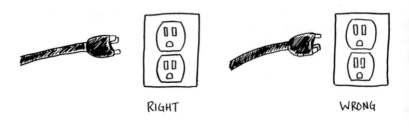

RIGHT   WRONG

— Every manual fills at least a page with *warnings*. One of the more hilarious is the Mr. Science lecture: "Electrical energy can perform many useful functions...." Which always ends with:

Do not pour liquid directly into your VCR.

(But if you sort of drip it in at an angle, that might be all right.)

7.   Think about Bill. <u>A True Story</u>:

A guy (call him "Bill")* purchases simple, low-priced VCR.
Bill takes VCR home.
Bill tries to make VCR work.
Bill can't.
Bill calls store.
Salesman (call him "Salesman") troubleshoots on phone.
No luck.
Bill comes down to store with VCR in brown paper bag.
Now it works fine.
Salesman suspects Pilot Error.
Salesman offers to go to Bill's house to see problem firsthand.
Bill declines.
More phone calls from Bill.
More store visits by Bill and Machine.
Then, nothing.
Salesman guesses: Either...

- Bill figured out VCR
- Bill threw VCR out window
- Bill threw himself out window
- all of the above

Soon after, Salesman reads newspaper article about local eye surgeon, pioneer in laser surgery techniques.... Guess who?

<u>So</u>: When you get frustrated with your VCR, just remember this heartwarming story of Dr. Bill, Laser Surgeon, who didn't get it either. It won't solve your VCR problems, but it might make you feel better.

*Actual name: William

2. HOW VCRs WORK

This chapter is to give you a general picture of how your VCR works.*
This ought to help your instincts for troubleshooting if you ever need
to (and it may shed some light on the reasons for the various buttons
you'll see in later chapters).

Your VCR is really just a combination of parts of three much more
familiar devices:

- A tape recorder   • A TV set   • A digital alarm clock

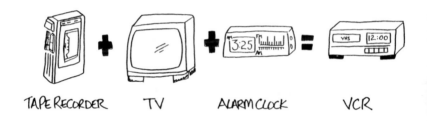

TAPE RECORDER   TV   ALARM CLOCK   VCR

---

*This is the *general* picture—don't sweat the details. If you're really into 29-micron head gaps,
heterodyne converters or other such technobabble, you don't need this chapter anyway. Go build
a time machine or something.

First of all, *what happens* when you put a tape into your VCR?

Open the front door of your VCR and look inside (or, if it's a top-loading model, look down through the top). See that silver drum, tilted at an angle? That's the **head drum** (the tape recorder part of the VCR).

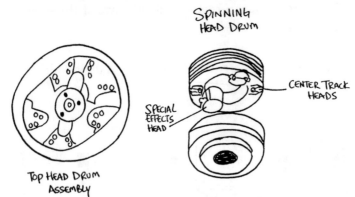

SPINNING
HEAD DRUM

CENTER TRACK
HEADS

SPECIAL
EFFECTS
HEAD

TOP HEAD DRUM
ASSEMBLY

Positioned around the head drum are the video **heads**, which do the actual videotape recording. Most VCRs now have either 2 or 4 heads. 4 heads are better than 2; they record a better picture and offer special effects—freeze frame, slow motion, etc.

When you put the video cassette tape into the machine, the VCR:

a. "swallows" the tape back into its interior
b. opens the front hinge on the cassette
c. wraps the video tape around some rollers and around the head drum—in an **M**-shaped pattern.*

*Beta VCRs work a little differently, but the principle is the same.

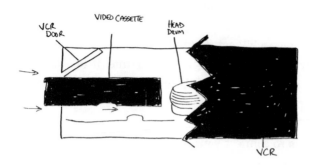

a.

VCR DOOR

VIDEO CASSETTE

HEAD DRUM

VCR

b.

VCR DOOR

VIDEO CASSETTE

CASSETTE HINGED DOOR

HEAD DRUM

VIDEO TAPE

VCR

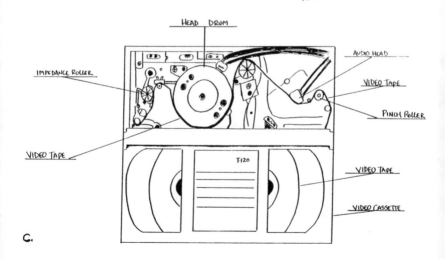

HEAD DRUM

AUDIO HEAD

IMPEDANCE ROLLER

VIDEO TAPE

PINCH ROLLER

VIDEO TAPE

T-120

VIDEO TAPE

VIDEO CASSETTE

c.

So much for the tape recorder part of your VCR.

Your VCR is also a TV set, *but without the picture tube.* Like a TV, it receives a stream of untuned signals and tunes in (selects) one channel from that stream. It just doesn't have a screen of its own to show that channel in picture form.

So the idea is to connect your antenna (or the cable from your cable company's antenna) directly to your *VCR.* Then, from there, your VCR can pass that same signal stream on to your TV:

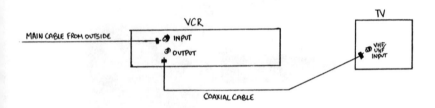

---

*Don't worry about the details and options here—you'll get a blow-by-blow account of how to hook up your VCR in the next chapter, cleverly titled "How to Hook Up Your VCR."

Thus, your VCR never records "from the TV"—a common misconception. The entire recording process happens inside the VCR. There, its tuner receives the signal stream from the antenna/cable, tunes in one channel, and records that channel directly into the tape recorder (the drum with its recording heads). And all this happens *before* the signal stream ever reaches your TV.

Your TV, by contrast, can only *display* (rather than *record*) the channel it selects from the signal stream. Of course, that stream may come directly from an antenna, *or* through your VCR (see above)—but either way, your TV tunes according to *its* channel setting.

The Moral Of The Story:

> The channel that your TV shows you is *not* necessarily the channel that your VCR records.
>
> Each device has its own tuner and can select a different channel from the same signal stream. The VCR simply receives that signal stream first—it's "upstream" from the TV (in fact, the VCR can record just fine *without any TV connected to it at all*).

So, how do you know when "what you see" on your TV is "what you get" recorded on your VCR?

Use the **TV/VCR** (or **TV/Video**) switch on your VCR.

When that switch is set to **VCR** or **Video** (there's usually a light to indicate this), you can watch the channel that your VCR is "seeing." If the switch is set the other way,* you can't:

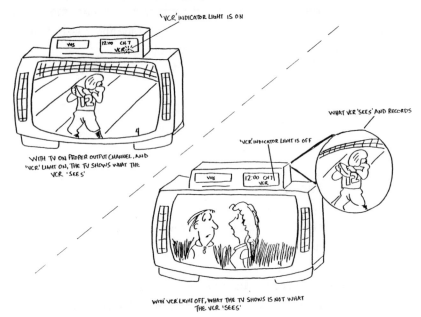

'VCR' INDICATOR LIGHT IS ON

WHAT VCR 'SEES' AND RECORDS

'VCR' INDICATOR LIGHT IS OFF

WITH TV ON PROPER OUTPUT CHANNEL, AND 'VCR' LIGHT ON, THE TV SHOWS WHAT THE VCR 'SEES'

WITH VCR LIGHT OFF, WHAT THE TV SHOWS IS NOT WHAT THE VCR 'SEES'

You may tend to overlook this if you use your VCR chiefly to play tapes. This is because most VCRs automatically switch to **VCR** (or **Video**) when playing a tape. But they *don't* do this when they're recording.

---

* Often there's no indicator light to tell you when the switch is set to **TV**. Such a light would be logical, so most manufacturers carefully avoid it.

Then there's the third familiar component of your VCR: A digital alarm clock . You can set your VCR to turn on and off according to its own clock, thus eliminating the need to be there to begin recording a TV show. You can record the show *in absentia** and watch it later at a more convenient time.

Naturally, this is the part that most captivated the public's imagination back in the 70's, when Sony Corporation dubbed its VCRs "Time Machines." The sales pitch: For the first time, you weren't a slave to the whims of the TV network programmers.

True. You were instead a slave to the whims of VCR designers.

The timer is the easiest to understand conceptually, but with all its options and jargon, it's also the easiest to screw up. But Fear Not: If you can set your digital alarm clock to wake you up in the morning, than you can set your VCR timer too (you'll learn how in Chapter 8).

*or wherever you happen to live

3. How to Hook Up Your VCR

Actually, you could just grab some neighborhood kid off his skateboard (Josh or Corey—all kids on skateboards are named Josh or Corey) and pay him twenty bucks to hook up your VCR for you. It will take him about five minutes. But if he's hired out already, read this chapter.

The idea is to hook up one device to another—in the correct order and at the correct orifices. To start, repeat this mantra several times aloud:

> **INPUTS** go to **OUTPUTS**.   **OUTPUTS** go to **INPUTS**.
> **INPUTS** never go to **INPUTS**.   **OUTPUTS** never go to **OUTPUTS**.

INPUT is where a device *receives* a signal; **OUTPUT** is where it *sends* a signal. You are hooking up devices to send and receive signals from one another.* Now, sooner or later, everybody grasps this. To counteract that, VCR makers decided not to agree on how they labeled the **INPUT** and **OUTPUT** connections. Thus, on your VCR, you may find

INPUT	and	**OUTPUT**
cleverly disguised as		masquerading as

CABLE IN	
ANTENNA IN	
ANTENNA INPUT	
IN FROM ANTENNA	**OUT TO TV**
RF INPUT	**RF OUTPUT**
VHF INPUT and **UHF INPUT**	**VHF OUTPUT** and **UHF OUTPUT**
or **VHF/UHF INPUT** (combined)	or **VHF/UHF OUTPUT** (combined)
TV IN	**TV OUT**
INPUT	**OUTPUT**

---

*Stupid Analogies: **(i)** If your job is to open the mail for your boss, then your Out Basket contents go into his/her In Basket; **(ii)** If you live downstream from a sewage treatment plant, then your In Pipe water comes from their Out Pipe water.

The Output Channel

To begin the Hooking-Up Procedure, here's the first thing to check: The **output channel** of your VCR is the channel over which it sends its tuned picture signal to your TV. *To view this signal on your TV, you must set the TV to that particular channel**—either channel *3* or *4*.

So, which is the output channel on your VCR—3 or 4
—and how do you tell?

There's a switch on your VCR (usually on the back—check the manual); you must *select* which one you will use.** But don't just select one at random. *It should be the channel that is not broadcast in your area.* If channel 3 is broadcast locally, use channel 4 as your VCR output channel, and vice versa. Otherwise your TV will be receiving a live broadcast over the airwaves on the same channel it is trying to receive your VCR's playback—and the broadcast will stomp all over the videotaped picture.

<u>Again</u>: To get a clear picture on your TV from your VCR, set your TV to your VCR's output channel. And that output channel (either 3 or 4) should be the channel with no local broadcast interference.

---

*This idea works similarly for the converter box provided with some cable TV services: The channel they tell you to turn your TV to is the *cable box's* output channel.

**With their usual thoughtfulness, manufacturers often label this switch with black letters on a black background—almost impossible to read.

After you know which channel to use as your VCR output channel, the next step is to connect your VCR to the signal stream coming from the Outside World.

Where is that signal stream coming from? Find out....

This is a test. For the next few paragraphs, this book will be conducting a test of the broadcast system in your area. This is *only* a test.

The Mr. Ed Test

Question: Can you watch "Mr. Ed" six or seven times a night if you want to?

Answer 1: Yes.
Conclusion 1: You must have cable TV; no normal TV station broadcasts "Mr. Ed" even once a night. Continue reading on the next page (Cable TV Hookups).

Answer 2: No.
Conclusion 2: You must have *non*-cable (broadcast) TV—signals taken from your own antenna. Skip ahead to page 34 (Antenna Hookups).

Answer 3: Yes—in several languages.
Conclusion 3: You must have a satellite dish—your own personal backyard cable company. Skip ahead to page 30, (option 2) and wherever it says "converter box," substitute "satellite dish receiver" instead.

Cable TV Hookups

If your televised signal stream comes from a cable company, there's a round *coaxial cable* (usually black) that comes in from the outside through a hole in the wall or the floor:*

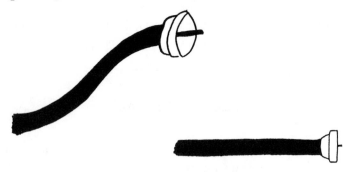

This cable supplies all of the channels that you get—whether you want them or not. Usually there's a set of channels for which you pay a flat monthly fee ("basic cable")—the 3 major networks, CNN, ESPN, etc.

Then, for even more money, you may also get one or more premium channels—HBO, Cinemax, The Movie Channel, Showtime, The Disney Channel, etc.** If so, then the device hooked directly to your cable may be a *descrambler*, or *converter box*, to decode these channels' signals so that people can't tap into cable lines to view them without paying.

---

*Technical name: "The cable that comes in from the outside through a hole in the wall or the floor."

**If it shows nudity, subtitles, or "Return of the Second Cousin of the Son of Flubber," it's a premium channel. These can be worthwhile, if you watch them enough to recover the savings on movie rentals (e.g. if your teenagers think that seeing "Ghost" 63 times a week is not excessive).

The need for scrambling and descrambling has complicated things. Basic cable stations come into your home clean (unscrambled) and ready for your VCR or TV to tune directly. If that were the case for all channels, life would be relatively simple.* But life is never relatively simple (relative to what, exactly?). Now you have three options....

Option 1

If you have just basic cable (i.e. no premium channels) *or* if you have premium channels but never wish to record from them (i.e. you'll record only from the basic cable channels), see page 28.

Option 2

If you have premium channels, and you want to be able to record from any channel (premium as well as basic), see page 30.

Option 3

If you have premium channels and want to be able to record from any channel and watch any other channel in the meantime, see page 32.

---

*You might consider nixing the premium channels altogether. Many of those programs can also be rented, so maybe you could put the money you're now spending on premium channels to better use simply by selectively renting the stuff you really want. Unfortunately, the cable companies are getting wise to this clever consumer ploy. Even if you don't subscribe to premium channels, some cable companies are scrambling even the basic cable signals—to *force* you to use their converter box. They figure that once you have a descrambler box in your house, it's more tempting for you to add premium channels or "Pay Per View" events (like "Super Wrestling Federated Tag Team Steel Cage Death Match of the Century, Held Monthly Until Somebody Is Killed, Which Is Pretty Unlikely Because Nobody Really Hits Anybody Too Much").

Option 1

A. Read the Advantages and Disadvantages of this setup (opposite, below) while you study the diagram above it. If it's what you want, continue with steps **B-D**.

B. Connect the cable from The Outside World to the VCR **VHF INPUT**.*

C. If you do receive premium channels—or if you have a TV that's not "cable ready,"—and therefore have a converter box, connect your VCR's **OUTPUT** to the converter box's **INPUT** (often labeled **CABLE**). If you have no converter box, skip this step.

D. Connect the converter box's (or, if you have no cable box, connect the VCR's) **OUTPUT** to your TV's **ANTENNA INPUT** (which may be called **VHF INPUT**, **VHF**, or **VHF/UHF INPUT**). On some older TVs, the **VHF INPUT** will not accept a coaxial cable. Instead the set has a connection for the older type of antenna wire, called "300-ohm," which looks like this:

If so, you'll need a 75-to-300-ohm transformer to adapt the coaxial cable connector for use on your TV (the cable company will provide one, if needed, when they install your cable):

*When you have cable TV, you have no need for **UHF INPUT**s or **OUTPUT**s—ignore them.

OPTION ONE:
COMPLETE DIAGRAM

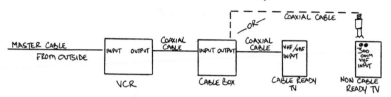

Option 1 Advantages:

- You can watch any channel at any time, even while your VCR records a different (unscrambled) channel.
- For recording in your absence, you can program your VCR to *change channels* (not possible if the VCR and converter box are connected in the reverse order—see Option 2).*

Option 1 Disadvantages:

- You can't record any premium channels this way (the signals from the premium channels aren't unscrambled until after they pass through the VCR to the converter box).

(Skip now to page 36.)

---

*Some cable companies offer a *programmable* converter box, so that it, too, can change channels when you're not around. In that case, you *can* record premium channels *in absentia*, if you're willing to program *two* timers ("yippee!")—one for the converter, one for your VCR. Or, you can get VCR Plus +, a little electronic device that does all the channel switching for you (more on this later).

Option 2

A. Read the Advantages and Disadvantages of this setup (opposite, below) while you study the diagram above it. If it's what you want, continue with steps **B-D**.

B. Connect the cable from The Outside World to the converter box (that's the descrambler) **INPUT**—often labeled **CABLE**.

C. With a coaxial cable, connect the converter box **OUTPUT** to the VCR's **VHF INPUT**.*

D. With another coaxial cable, connect the VCR's **OUTPUT** to the television's antenna **INPUT** (which may be called **VHF INPUT**, **ANTENNA INPUT**, **VHF**, or **VHF/UHF INPUT**). On some older TVs, the **VHF INPUT** will not accept a coaxial cable. Instead the set has a connection for the older type of antenna wire, called "300-ohm," which looks like this:

If so, you'll need a 75-to-300-ohm transformer to adapt the coaxial cable connector for use on your TV (the cable company will provide one, if needed, when they install your cable):

*When you have cable TV, you have no need for **UHF INPUT**s or **OUTPUT**s—ignore them.

OPTION TWO:
COMPLETE DIAGRAM

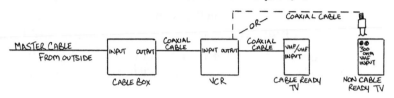

Option 2 Advantages:

- You can watch any channel at any time.
- You can record any channel—even a premium channel.

Option 2 Disadvantages:

- You can watch only the station selected by the converter box (in this setup, it's "upstream" from everything else; since it outputs just one channel at a time, that's all you'll see "downstream").
- You can't watch one station while recording another.
- Not too common: If your converter box uses channel *2* as its output channel (recall the discussion on page 24), you'll have to switch the TV channel to 2 (rather than leaving it on the VCR's output channel—3 or 4) when you're not using the VCR.

(Skip now to page 36.)

Option 3

A. Read the Advantages and Disadvantages of this setup and study the diagram. *Note: Each connection must be made with a separate coaxial cable.* If it's right for you, do steps **B-D**.

B. Connect the cable from The Outside World to your antenna splitter's **INPUT**. Connect the two antenna splitter **OUTPUT**s, respectively, to: the **INPUT** (often labeled **CABLE**) of converter box #1 (the descrambler); and the **INPUT** of converter box #2.

C. Connect the **OUTPUT** of converter box #1 to the VCR's **VHF INPUT**.*

D. Connect the VCR's **OUTPUT** to **INPUT A** of an antenna A/B switch.** Connect the **OUTPUT** of converter box #2 to **INPUT B** of the A/B switch. Connect the **OUTPUT** of the antenna A/B switch to your TV's **ANTENNA INPUT** (which may be called **VHF INPUT, VHF,** or **VHF/UHF INPUT**). On some older TVs, the **VHF INPUT** will not accept a coaxial cable. Instead the set accepts an older type of antenna wire, called "300-ohm," which looks like this:

If so, you'll need a 75-to-300-ohm transformer to adapt the coaxial cable connector for use on your TV (the cable company will provide one, if needed, when they install your cable):

*If you have cable TV, you have no need for **UHF INPUT**s or **OUTPUT**s—ignore them.

**Some of the newer TV/monitors have built-in A/B switches. If yours does, use it—don't buy one.

OPTION THREE:
COMPLETE DIAGRAM

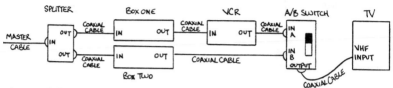

Option 3 Advantages:

- You can watch any channel at any time.

- You can record any (even a premium) channel: Set the A/B switch to A; set converter box #1 to the channel to be recorded. Set your VCR to the cable box's output channel (2, 3 or 4); set your TV to the VCR's output channel (3 or 4). Start recording.

- You can watch any channel while recording any other: Follow the above recording instructions, then set the A/B switch to B and use converter box #2 to watch whatever else you want. At any time, you can set the A/B switch back to A to check on your recording.

Option 3 Disadvantages:

- As with option 2, if you're recording when you're gone, your VCR is still limited to the channel you've preselected on your converter box (unless you're using VCR Plus+).

- This setup gets a bit expensive with all the extra paraphernalia.

- Your TV room starts to look like Frankenstein's laboratory.

(Skip now to page 36.)

Antenna Hookups

If you don't have cable, you probably have an antenna of some sort. There are two basic types of antennas: outdoor (up there on your roof) or the indoor "rabbit ears" antenna that sits on or near your TV.

Outdoor Antenna

If you have one of these, it most likely uses the same kind of coaxial cable that the cable companies use—called "75-ohm" cable. Coaxial cable has pretty much replaced the older style flat wire, also called "300-ohm" or "twin lead." *

Connecting any antenna to a VCR is relatively simple, and (for once), the manufacturers' Instruction Manuals for your VCR (hope you haven't CRATTIBIPped them yet) and for your antenna are the best places to read exactly how it's done. Here's the general idea:

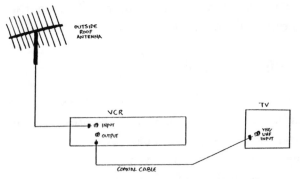

*Without getting into a big technical harangue about it: 75-ohm cable is the a superior method of delivering a TV signal because it has "lower signal loss" than the older, 300-ohm wire. This means you get a clearer, less noisy picture.

Indoor Antenna (Rabbit Ears)

The rabbit ears that come with and/or are built into your TV are not the best for hooking to your VCR.

First of all, the built-in kind are free, and as you know, you get what you pay for. Secondly, the wires the TV manufacturers use to connect those rabbit ears to the TV are usually just long enough to reach the proper antenna input(s) on the TV. They don't make allowances for VCRs. There are ways to extend those wires, but it's not recommended.

The best solution in these situations is to spend an extra $20 for a stand-alone rabbit-ears antenna with its own base—so that you can move it around and install it closer to the VCR. This will give you much better reception. Yes, it's a little more money, but remember that your VCR recordings can be no better than the quality of the original signal; that extra $20 represents a lot of value.

Again, follow the manufacturer's Instruction Manual guidelines for connecting the antenna to your VCR. This is the basic idea:

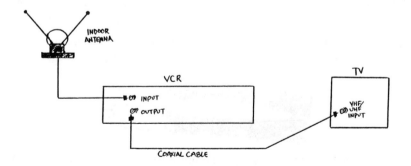

Audio/Video Monitor Connections

The other considerations in hooking up your VCR are your Audio/Video connections. These are for *playback* only; it doesn't matter whether you have an antenna, satellite dish, or cable service.

If you have an older TV, just skip to page 40. But if you have a newer TV, it's probably a *monitor/receiver*. That means simply that your TV is not just a television (the "receiver" part) but also a video monitor, which can display pictures from VCRs, camcorders, computers, etc. Here is the back of a typical monitor/receiver.

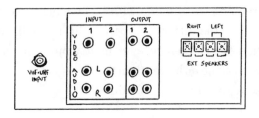

You see the familiar **VHF/UHF** antenna inputs, plus some other jacks, called **AUDIO/VIDEO** (or **A/V**) **INPUTS**. These can deliver superior picture and sound signals from your VCR. You connect the **AUDIO/VIDEO OUTPUTS** on your VCR to the corresponding **INPUTS** on the back of your monitor ("**OUTPUTS** go to **INPUTS**, blah, blah, blah"). The plugs you use for these connections are called "RCA phono plugs:"

RCA invented them, so they got to name them. But *any brand* of TV and VCR may use them now (as do cassette decks, CD players, etc.).

If you own a monitor, you really should take advantage of the better picture and sound it offers (you'll find RCA plugs just about anywhere electronic products are sold). If you're not sure what to ask for, bring this book or your instruction manual and show the salesperson what you want to do. The cables will cost you less than $20 and provide a marked improvement in both picture and sound.

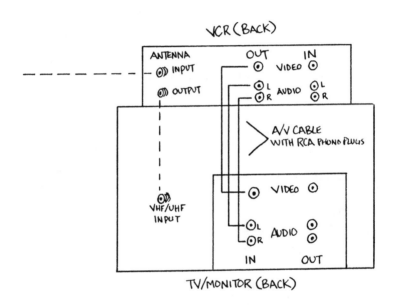

VCR (BACK)

A/V CABLE WITH RCA PHONO PLUGS

TV/MONITOR (BACK)

Furthermore, if you have a stereo VCR, you will *not* hear the tapes in *stereo sound* unless you connect the VCR to the monitor's proper **INPUT** jacks (or to your stereo system, as shown on the next page). Just using your VCR's output channel to play tapes is fine for the picture, but you won't hear the stereo audio separation.

Question: You own a nice HiFi Stereo VCR, but not a monitor TV. Do you have to go to the extra expense of a monitor?

Answer: Nope. You can connect your stereo VCR to your stereo music system and use it just like any other stereo audio component, such as a turntable, cassette deck, or compact disc player! And this is not just for hearing the stereo soundtracks on rented movies: If your VCR has MTS (Multichannel Television Sound—otherwise known as Stereo TV), its tuner can receive stereo television signals right over the air—and you can watch the picture on your mono TV while hearing stereo sound through your VCR and your own stereo music system!

How? Like this:

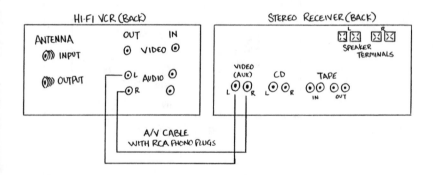

Connect just the **AUDIO OUTPUT**s of your VCR to any **AUXILIARY INPUT**s on your stereo receiver (generally, the newer model stereo receivers have inputs labeled **TV** or **VIDEO**; older models may just be labeled **AUX**). In fact, you can use any input on the back of the receiver except for the **PHONO INPUT**.

Of course, if you do have a monitor TV to go with your HiFi Stereo VCR, you've got the makings of a full-blown A/V (Audio/Video) system— taking full advantage of the HiFi soundtrack of your VCR, the better picture from the monitor, and the better sound of your stereo system:

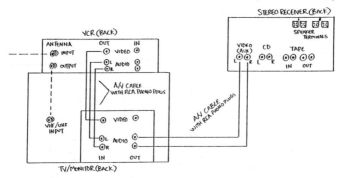

You connect the VCR's **OUTPUT**s to the monitor's **INPUT**s. Then you connect the monitor's **AUDIO OUTPUT**s to your receiver's **AUXILIARY INPUT**s, as outlined on the bottom of page 38.

And then, for a little *more* money, you can turn your viewing room into a true home theatre (large screen, four-speaker surround-sound, popcorn incense, sticky floors).... It's all up to you and your banker.

4. HOW TO TUNE YOUR VCR

<u>Before you start this chapter</u>: If you subscribe to premium channels *and* you've hooked up your VCR using last chapter's cable option 2, cable option 3, or the satellite dish configuration, then you can skip this chapter entirely; you don't need to do any tuning, because your VCR will always be set to the output channel of the cable converter box or satellite dish receiver.

However, if you decided to use cable option 1 (or have antenna service only), then your VCR needs to be tuned properly....

If your VCR has a ***quartz tuner*** (or if you can catch Josh or Corey between gigs), you still don't have much to worry about. Just like a video-smart skateboard kid, a quartz tuner searches for and locks in each station from the cable's or antenna's signal stream, essentially eliminating the need for you to manually locate your stations.

In fact, some quartz tuners will even tune the frequencies of scrambled stations*; some will skip over these. But most quartz tuners allow you to eliminate unwanted stations (like the scrambled ones or other stations you rarely watch) by skipping them, but you don't have to if you don't want to bother.

So if your VCR has a quartz tuner, just look in your instruction manual for the one instruction that tells you how to *activate* this tuner. After that, it's automatic.

---

*And, sure enough, that's what you'll see on the TV screen—a scrambled signal (no tuner will *unscramble* the signal).

OK, so what if you *do* use Hookup Option 1 but you *don't* have the latest-and-greatest quartz-tuned VCR? You're in good company. Millions of VCR owners have machines that were built with an evil device called a ***varactor tuner***. Back in the Reagan Years, *most* VCRs (and many TVs) were infested with varactor tuners. Indeed, Dr. Bill, the laser surgeon of Chapter 1 fame, was the victim of acute varactoritis.

Remember the push-button car radio? It had five buttons—presets you could use to tune to your favorite five stations without taking your eyes off the road for long. To preset these stations, you'd manually dial-tune each station, then pull-and-push the button you were designating for that station. This is basically how a varactor tuner works in your VCR.

The most common method of hiding the varactor tuner is a little door, either in front of or on top of the VCR. Inside is a series of switches with little dials next to each switch. And on the front of the VCR is a series of windows that show channel numbers and maybe some asterisks (*):

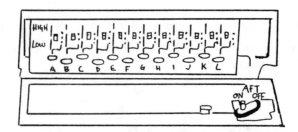

Each little switch and dial is, in effect, an individual TV tuner. With it, you can dial in any station you want—just as you can set each of the five push-buttons on your car radio to the same station.*

The point is, all varactor VCRs share certain common traits:

- Stations 2–13 were pretuned at the factory, but no stations above 13 are pretuned. You have to locate and store (and which of the stations below 13 are unnecessary and can be deleted).

- The manuals generally give you lousy instructions on how to do this. For example, *none* of them say this:

```
ATTENTION ALL VCR OWNERS:  IF YOU'RE TRYING TO
TUNE IN CABLE STATIONS ON YOUR VCR, IT MIGHT HELP
YOU TO KNOW THAT THEY ARE NOT IN NUMERICAL ORDER.
CHANNELS 14 THROUGH 22 ARE ALL SORT OF SCRUNCHED
IN THERE BETWEEN 6 AND 7.  DON'T BLAME US—BLAME
THE F.C.C.
```

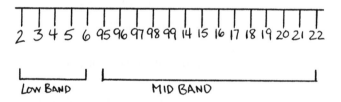

Yes, Virginia, this really is how the station numbers work.

---

*Other varactor tuner setups look different but work essentially the same way. For instance, instead of fourteen individual switches and dials, maybe your VCR has one dial and a memory button—the idea being that you dial in each station, push a button to tell your VCR to "memorize" that setting, and then you go to the next position.

So, before you actually get down to the mechanics of tuning in stations, look at the following chart, which has been lovingly and painstakingly created to show you the exact order of all of the stations as they appear on cable.

You probably don't get all of the stations listed here, but every station you do get will be on here somewhere:

THE COMPLETE VHF AND CABLE TV TUNING GUIDE

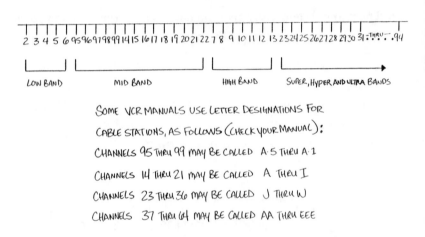

Yes, this does look like an FM radio band—for two very good reasons:

- Most people are familiar with FM radios and how to tune them.*
- The FM radio band is right smack in the middle of the same VHF (Very High Frequency) band that carries most of the commercial TV stations and all of the cable stations.

*Peculiar thing about FM radio: Call 'em crazy, but all of the stations are in *numerical order* (is that wacky, or what?)! If you're looking for a station at 99.5, you *know* it'll be between 99.3 and 99.7.

Use this chart when you're trying to tune in your VCR (examples to follow). That way, if you've just set, say, channel 22, and 23 is the next one you want to find, you'll know that it's *not* next to 22—it's next to 13.

A couple more notes before you practice with a few examples:

All *cable* stations are on the VHF band. Despite what you think, *none* of them are on the UHF band. This is a very common (and understandable error) people make—looking for stations above channel 13 on the UHF band. It's like looking for an AM radio station on the FM dial.

However, you will note on the chart that the VHF band is divided up into "low" and "high" stations. Some of the older VCRs have divided up their tuners accordingly, and—to complicate things still further—they sometimes designate channels above 22 as the **CATV band**. Don't sweat the name; just look on the chart and it will spell it all out for you.

OK, here's how to tune in your stations (before you start this process, remember that your VCR must be hooked up like cable option 1 from last chapter; if you've decided to use cable option 2 or 3, or if you have a satellite dish, there's no need to fine tune your VCR at all, and you needn't bother reading this chapter)

*Remember to set the **TV/VCR** switch to **VCR**, so that your TV screen will show what your VCR is "seeing."*

1. Determine how many stations your VCR can remember at a time (check the manual). Clues: If there are 14 dials, you can fit 14 stations. Or, if your manual says it has "sixteen channel position memory," then you fit 16 stations.

2. Write down, horizontally and in numerical order, the cable stations you want to preset—up to your VCR's preset limit. For example, your cable system may carry, say, channels 2 through 42, from which your VCR can preset up to sixteen.

3. Underneath that list, write down, horizontally, the stations that are *currently* stored in the VCR.* Now you have a list of the stations you want, and a corresponding list of those you have. By using the cable station chart as a guide, you can easily go from the station you're on now to the one you need.

---

*As you go through the stations, either check your cable guide, wait for station identifications, or have a Helper watch a second TV and match stations with you by comparing descriptions of what's on the screen:

You: "OK, I got a guy with waist-long hair, screaming and pretending to play a guitar while being assaulted by nearly-naked women.

Helper: "OK, that must be C-SPAN Senate Hearings... no—wait... I guess it's MTV."

When tuning: **(i)** select the right band—remember the AM vs. FM radio analogy; **(ii)** find where on this band you are at the moment; and **(iii)** move to where you want to be.

Example: You've got channel 4 preset, but you want to change that preset to channel 6.

Look at the chart. Both stations are on the low band, so stay on the low band setting and just tune up (in frequency): 4... 5... 6—just like changing a radio station. Lock in this setting *as directed in your owner's manual.*

Example: You've got channel 7 preset; you want to change it to 16.

The chart tells you to go *backwards* (tune *down*): 7... 22... 21... 20... 19... 18... 17... 16. Lock in this setting.

Example: You've got channel 5 preset; you want to change it to 25.

Look at the chart. Channel 5 is on the low band; channel 25 is on either the high band or the CATV band, (depending on how your VCR's tuner shows them). Change the band switch on the position marked channel 5 from **LOW** to **CATV** (also called **SUPERBAND**)....

Channel 5 disappears; what you see now—if anything— is a location on the CATV band. Find an identifiable station on that band—say it's channel 36. Now it's easy from the chart to see that you should tune backwards from 36: 35 ... 34 ... 33 ... etc., to 25. Lock in this setting.

Now tune *your* VCR. Keep the manual handy to tell which button(s) to press for locking in, etc. In case of emergency, call Josh or Corey.

5. How To Play a Tape

You probably don't need much help with this topic—most people have rented a movie now and then. You just stick the rewound tape into the VCR, press **PLAY**, and grab the popcorn, right?

Right.

But just in case you have any problems, here's a list of things to check when trouble shooting.

- Check all of the physical connections—make sure they're all tight (and not accidently reversed).

- <u>Check the channel settings</u>. Go back to page 24 and reread about output channels. If you push **PLAY** and don't get a clear picture, check to be sure that your TV (and cable converter box, if any) is set to the proper output channel (also, you might double-check to make sure that the 3-4 switch on the back of your VCR hasn't accidently been moved).

- <u>Check the **TV/VCR** mode</u>. If you're sure everything is on the right channel, it could be that you have an older VCR that doesn't automatically switch to **VCR** (or **VIDEO**) mode when playing a tape. There should be a visible light that says **VCR** or **VIDEO**, or a red or green light, or *something*. Check your manual.

- <u>Check the tracking</u>. If you have a picture with lines or fuzz, usually a simple adjustment of your tracking control* will fix it. Adjust the tracking, back and forth, until the picture looks best. The **TRACKING** control is usually a small rotary knob on your VCR, often tucked behind a door.** It's clearly marked and (believe it or not) every manufacturer calls it "tracking." 95% of the time, tracking adjustments will clear up any picture problems (and you can't hurt anything by trying it, so try it first).

- About 5% of the time, you'll have either a <u>defective tape</u> or <u>dirty heads</u> (remember the head drum?). To determine which, first try another tape or two. If the problem goes away, obviously, it was the tape; if it doesn't, you may need to clean those video heads.

  Now don't go squirting any shampoo in there. Either use a head-cleaning cassette designed especially for the purpose, or (better yet) take the VCR to a qualified repair center and pay them to give it a professional going over. It's well worth the cost ($24-36); if the heads get too dirty, they can become permanently damaged and need replacement—and you don't even want to know how much that costs. Preventative VCR maintenance costs a little but will often save you big bucks later.

*Don't confuse the tracking control with the "sharpness" control, which increases or decreases the amount of detail in the tape's picture.

**On some newer VCRs, the tracking adjustment is located on the remote control, and on many higher-priced machines, a feature called "digital tracking" makes the adjustment automatically.

- <u>Flagging</u>. If the top of the picture "bends" to the left or right (called "flagging"), it could be a very old TV, a height adjustment on your TV, an out-of-shape VCR, or a bad tape. Try adjusting the **HEIGHT** control, if you have one on your TV, or try a different tape. Otherwise, you probably need to have it serviced.

- <u>No sound?</u> It's probably a bad tape (always suspect the tape first whenever a new problem suddenly appears), or a bad audio head.

- <u>Weird colors?</u> If the colors are too rich—they "bleed"—turn down the color control on your TV. If that doesn't help, you probably just have a poorly recorded or cheap videocassette.

- <u>If the picture rolls up or down</u>, adjust the **HORIZONTAL** control.

- <u>If all else fails</u>, maybe it's your TV—try the VCR on a different television.

If *none* of this works, and you're pretty sure it's not just bad karma,* then you've done all that you can, and you should take it to the shop. But try *all* of these troubleshooting suggestions first. It's very frustrating to be charged $35 to have the technician tell you "no trouble found."

---

*Actually, some of the full-service shops seem to be able to cure even bad karma.

6. HOW TO RECORD A TV SHOW WHILE YOU WATCH IT

Voilá—Your VCR is hooked up and tuned in, and you can play a tape with the greatest of ease. And you can go for years knowing very little more than this about your VCR. But after a while, the luster of renting movies may wear a little thin.

At first, your local video outlet is like a candy store (with 63 million movies—how can you miss?) But then you start to see why there are so many titles in the store—nobody ever checks out most of them. Everybody wants to rent the same five movies you do:*

So unless your tastes run to the classics ("Beach Blanket Bingo") or to art films ("The Unseeable Lighting of Bareness"), sooner or later you're going to get sick of competing with six hundred other customers— because 595 of you won't get what you want.

The solution? **Record!** Start earning a return on your VCR investment. Even if you don't have cable, there's some very worthwhile stuff on TV these days.**

Of course, the good programming is on when you're sound asleep or not home, so you have to use your timer to catch it—a subject covered in gory detail in chapter 8. But for starters, here are the mechanics of actually putting a tape into the machine and making a recording.

---

*The blockbusters that were just in the theater, but since your loan officer was on vacation, you never saw them: "Kindergarten Terminator"
"Die Even Harder: Why Is It, Everywhere I Go I Run Into Terrorists?"
"Sleeping With a Pretty Mystic Young Enemy Who's Dying for a Pizza"
"Dances with Gloves"
"Rocky MCLXVIII: The Next Final Challenge"

**Sure, there's a lot of dreck, but there are also programs like the "Civil War" miniseries.

- Turn everything on (VCR, TV and converter box, if any)

- Put a blank cassette into the VCR. Be sure that the record tab is intact on the front edge of the tape. If not, you can't record on the tape unless you cover the tab hole with a piece of tape:

- Select the recording speed, based on the length and desired quality of the recording. On VHS machines, the three possible speeds are: **SP** (Standard Play); **LP** (Long Play); **EP** (Extended Play—also called **SLP**, for Super Long Play*).

SP gives a maximum of 2 hours on a standard-length VHS T-120 tape; **LP** gives 4 hours;** and **EP/SLP** gives 6 hours, which is fine for most general purpose record-and-erase-later stuff. On newer VCRs, with HQ (High Quality) circuitry and/or four heads, the **EP/SLP** picture is quite adequate, but **SP** gives the best picture and sound. For important videos to be kept for years, use SP.

On Beta machines, the three speeds are Beta I, Beta II and Beta III. Only the oldest Beta VCRs used Beta I, which gives a maximum of 1.5 hours of recording time on a standard L-750 Beta tape. Beta II gives 3 hours; Beta III gives 4.5 hours.

---

*Coming soon: Really, Really, Really Long and Economical Play For Tightwads (RRRLEPFT).

On most VHS machines made after 1987, the LP recording speed has been phased out (although all machines will *play* LP tapes). And with LP, special effects like **PAUSE, **SLOW MOTION**, and **FAST FORWARD/ FAST REWIND SEARCH** may not work, so if SP isn't practical, use EP.

If you are using cable option 1 hookup or an antenna:

- If you don't use a converter box, set your TV to the output channel of your VCR (3 or 4). If you do have a converter box, set your TV on its output channel (2, 3 or 4) and set the converter box to the output channel of your VCR (3 or 4).

- Set your VCR to the channel you want to record.

If you are using cable hookup option 2, 3 or a satellite dish:

- Set your TV to the output channel of your VCR (3 or 4). Set your VCR to the output channel of your converter box or satellite receiver (2, 3, or 4.)

- Set your converter box or satellite receiver to the channel that you wish to record.

Everybody:

- Be sure that the **VCR** or **VIDEO** indicator is lit on the front of your VCR (most—but not all—VCRs go into this mode when you turn them on), so that the picture on your TV screen is the same as what the VCR is "seeing" (recall page 20).

- Push **RECORD** (or **RECORD** and **PLAY** together—see the manual). Some sort of **RECORD** indicator will show. You're now recording.

Strongly recommended suggestion: *Don't* wait for a once-in-a-lifetime program to try this for the first time. Try it with "America's Goriest Home Videos" or something—for just a minute or so. Then stop the tape, rewind and replay, to check your success. If you didn't get it right, read through these instructions and try again.

7. HOW TO RECORD ONE TV SHOW WHILE WATCHING ANOTHER

That torn, desperate feeling—it can strike without warning:

It's Super Bowl Sunday, with a houseful of guests watching the pre-game show (now in its third day), with its clean, incisive analysis from the Has-Beens with the mikes.*

Suddenly you realize that the final episode of a terrific 16-part, made-for-TV miniseries** will be starting on channel 7, at about the time of the second-half kickoff of the Super Bowl (channel 10).

What do you do?

A. Call channel 7 and ask them to delay the movie.

B. Call the Super Bowl and ask them to take a 2-hour halftime.

C. Record the movie on channel 7 while you watch the rest of the game on channel 10.

D. Make one of life's Big Decisions.

Unless you work for the State Department, options **A** and **B** are pretty well out. Of course, option **D** is unthinkable. Fortunately, **C** is not all that tough....

---

* *Dan:* "Well, Bob, Ditka is a Capricorn with Aquarius rising. That should give him the advantage here on the artificial turf."

  Bob: "Yes, Dan, but Shula's record in even-numbered years—especially when he wears red socks—is really impressive. Look for that to help his special teams."

** "The History of the Universe, at Least as it Pertains to Money, Sex, and Power, Anyway"

That is, it's not that tough as long as you haven't hooked up your VCR according to cable option 2 (pages 30-31).

If you did hook up option 2, you simply cannot record the movie and watch the Super Bowl at the same time. That's because under option 2, the converter box's tuner is the only one that counts—it's "upstream" from everything else. So for your Super Bowl Sunday dilemma, you'd simply have to make a choice: Set your converter box on channel 10 and watch the rest of the game; or set the converter to channel 7 and see the melodrama instead.

But, if you're using cable option 1 or an antenna, you have *two* tuners that you can employ—the tuner built into your VCR, and the tuner in the television or converter box.

When the **VCR** indicator is showing, the only signal you can see on your TV screen is coming from the VCR.

Push the **TV/VIDEO** button.

Once you push that button and the **VCR** light goes out, you shut off the VCR picture's path to the TV screen. Then, no matter what your VCR is doing—whether or not it's recording (in this case channel 7)—the picture you now see on your TV is not coming from the VCR but from the *other tuner*—that in your TV or converter box.

So here's the solution to your Super Bowl Sunday dilemma (again, this works as long as you're using cable hookup option 1):

- Start your VCR recording on channel 7 (see chapter 6 for the step-by-step instructions as to how to record).

- Push the **TV/VIDEO** button so that the **VIDEO** indicator is *off*.

- Change your TV (or converter box) to channel 10.

That's it—no muss, no fuss—and not much time taken to do it, *once you know how.* So try it now, when there's nothing important to miss if you botch it the first time.

8. How to Record a Show
When You're Not Even There

Here's the love-hate relationship with your VCR: You love it when you rent movies. You hate it when you need to set the programmable timer. When you do finally master that programmable timer, though, your VCR will really start to earn its keep. Essentially, it lets you program your own personal broadcast network—a real boon, for inevitably the better the program, the less convenient is its broadcast time.

Unfortunately, setting your VCR timer correctly is an exact science: if all of the steps aren't exactly right, the recording isn't anywhere *near* right (instead of "Masterpiece Theater" you get "Let's Make a Deal"). For that reason, you do need to have your own machine's instruction manual handy here. Every VCR model has different buttons for the same things, and you simply must know the buttons for your machine. Then you can follow along here through a translated, explained version of the basic steps involved in setting your timer.*

---

*Note: It's now possible to buy your way out of needing this chapter, with a device called VCR Plus+ (at this writing about $60)—a handheld, remote-control device that, once properly programmed, will turn on your VCR (and your cable box, if necessary) at the right time, select the proper channel(s), make the recording(s), and then shut off everything. All you need to do is look up the desired TV show in your newspaper's listings (or in TV Guide magazine) and use the VCR Plus+ numeric keypad to punch in the code for that show (example: *Murphy Brown*: 90962).

There are some slight drawbacks to VCR Plus+: Each show's code must be entered individually. You can't simply tell it to record all 4 hours of daytime soaps as one program. VCR Plus+ recognizes TV *shows*, not blocks of time. This can be a problem, too, for recording live sporting events, which don't always end on the half-hour. A baseball game's code, for example, will tell VCR Plus+ simply to "record the show that starts at 1:30." That show (the ball game) is *scheduled* to end at 4:30 (and VCR Plus+ knows this), but if the Mets stage a ninth-inning rally, it'll get snuffed—not by Ron Dibble, but by the 4:30 scheduled ending time, obeyed by VCR Plus+ (yes, you can add time increments, but it's easy to forget to do this).

Ironically, if you can set up VCR Plus+ in the first place, you probably don't need it, for although its instruction manual is clear enough, the initial procedure is still quite involved. This makes VCR Plus+ a great gift idea—as long as the giver does that initial programming before wrapping it up.

These are the seven basic steps for programming your VCR's timer. *But*, depending on the machine you own, there may be up to nine steps for your machine. Furthermore—thanks once again to the thoughtfulness of competing manufacturers—the steps are *not* in the same order on every VCR (some machines want you to set the time first, others the channel, others, the day). Keep your manual handy as you read here!

0. Set your VCR's clock and calendar. Just like a clock radio, your VCR can't turn on at the right time if it doesn't *know* the right time (and date) in the first place. When setting your VCR's clock/calendar, you key in the current time, day, month and year; or (on older VCRs) you key in the current day of the week. See your instruction manual for the exact keystrokes. This step is numbered **0**, because you need to do it only once— not for every time you want to record a program (*unless the electricity fails*—one of Life's Little Frustrations).

1. Set your VCR to programming mode. It's simply not enough to say in a loud, commanding voice: "OK, VCR, I'm going to program you now." You need to press a button to bring up the proper programming **display**, either on your TV (if you have on-screen programming), or on the VCR's front panel.

2. Select the event number (also called the program number). Unless it's a very old, single-program machine, your VCR can be programmed with instructions for anywhere from two to nine different events. Each event is given a complete set of

instructions for recording a particular channel on a certain day and time. And in many VCRs those instructions can include a command to *repeat* the recording every day or every week. That way, you can program just one event (instead of five) to record "All My Shining, Restless Days in the Hospital" every afternoon at its usual time.

3. Set the proper date (or weekday) for that event. You key in the month and day your show is scheduled. Or—if your VCR knows only the day of the week—key in that information. For example, if today is Sunday, January 6th, and you want to record a show on Thursday, January 17th, you indicate "the second Thursday." Check your manual for the keystrokes.

4. Set the proper starting time for the event. Most newer VCRs use a digital entry method, where you punch in the time you want, digit by digit. For example, for a show starting at 8:00, you'd press ⓪, ⑧, ⓪, ⓪.* The display would read **08:00** Then you select AM or PM. Usually (but not always—check your manual) you press ① for AM or ② for PM.

Older VCRs use an "alarm clock" approach, where you push a button to scan through the hours and minutes—much as you would set an alarm clock. Some VCRs let you set the hours and the minutes independently; others make you scan through by minute only. And most let you go backwards, too, in case you're a little slow in letting up on that scan button.

*That first ⓪ is important, by the way. If you forget, the VCR thinks you're trying to set something to start at "80 o'clock" which isn't usually a valid time (except for important deadlines).

5. Set the proper time to stop the recording. Some VCRs, usually the "digital entry" types, ask you to key in a "stop" time just like the "start" time. For example, if a show ends at 9:00 PM, you press ⓪⑨⓪⓪, and select PM (maybe by pressing ②).

On the older, alarm-clock-type timers, you indicate the end time, either by scanning to that time (just as you scanned to the starting time), or by telling the VCR how long the event is. In this latter case, then, rather than telling the machine to start at 8:00 PM and stop at 9:00 PM, you'd tell it to start at 8:00 PM and record for one hour. Check your manual.

Note: Don't overlap your events—don't let them even "touch" each other. For example, do not program one event to record from 3:00 to 4:00 PM and another event to record from 4:00 to 5:00 PM. It takes your VCR a few seconds to finish an event completely. If the stopping procedure is set to begin at *exactly* 4:00 PM, it won't completely stop and be ready to "check its watch" for the next starting time until a few seconds *after* 4:00. *Thus it will miss the event that begins at exactly 4:00.*

The solution? Set the first event to record from 3:00 to *3:59*; set the second to record from 4:00 to 5:00.

6. <u>Set the proper channel to be recorded</u>. Don't forget this! All too often, the VCR comes on and records all right, but it's not the show you want—because you've forgotten to tell the VCR which channel to record (in which case it usually records the channel it was on at the event's starting time).*

Once again, the digital entry VCRs are easier:** Most won't let you proceed without explicitly specifying a channel. The older "alarm clock" type machines are trickier: They'll let you set the timer without selecting a channel.

7. <u>Set the recording speed</u>. Sometimes this is part of the timer procedure (steps **4** and **5** here)—especially with newer digital entry machines. You simply select a number (①, ② or ③) to indicate **SP, LP**, or **EP/SLP**. Or, on other machines, you push a **RECORDING SPEED** button.

With older VCRs, you must adjust a mechanical switch before you put the VCR into the "timer ready" mode. If you don't set this switch appropriately for the length of the event (and your available blank tape), you may not have enough tape. A three-hour football game simply won't fit onto a VHS T-120 tape if you're recording at SP speed.

---

*With cable options 2 and 3 (pp. 30-33), the channel to program into your VCR's event instructions is *the output channel of your cable converter box.* You then must leave your converter box **ON** and set to the station you're recording. For example, to record channel 10, you'd set your VCR to record channel 2, 3, or 4 (the converter box's output channel) and *leave the converter box on channel 10.*

**You may have noticed a definite pattern emerging throughout this book: Newer VCRs tend to be more "user-friendly;" older VCRs tend to be more "user-homicidal."

Those are the seven basic steps—but be sure to check them against the Order Of Doing Things listed in your instruction manual.

To practice, try the sample program(s) in that manual. Or, set the timer to come on ten minutes from the time you set it. If you're practicing at three PM, have it come on at 3:10, for just a couple of minutes. This way, you'll know instantly if it worked. If not, try it again, and check this troubleshooting list of the most common errors:

- <u>You forgot to push the "timer" button to "arm" the program to start automatically</u>—like forgetting to put your clock radio in the "alarm" position. Some of the newer VCRs go into this timer mode automatically when you turn them off; some don't.

- <u>You forgot to put a tape into the VCR</u> (don't laugh—this happens more often than you might think*). Most VCRs will alert you; some won't.

- <u>AM/PM Mistake</u>. You set an event's time for AM instead of PM or vice versa. Or, sometimes the VCR's *clock* (not the event's starting/stopping times) is incorrectly set to AM instead of PM, or vice-versa. If all your events are being recorded exactly 12 hours off, check the clock.

*even to (ahem) authors of VCR books

- <u>Wrong week and/or date</u>. Either your clock or your event times are incorrectly set—*both* must be right. If you set your event time perfectly for a Friday night recording, but your VCR's calendar is a day off, guess what doesn't happen?

- <u>Wrong channel</u>. Maybe you've mis-keyed something. For example, check to be sure that you haven't entered channel 6 as **60**.

  And all you users of cable hookup option 2 or 3, or satellite dishes, <u>remember</u>: It's the *converter box* (or satellite receiver) that you must leave (turned on) set to the show's cable channel; and in the VCR's event instructions, you must specify the converter box's (or satellite receiver's) output channel. The diagram on page 31 shows you why.

- <u>Overlapping programs</u>*. If you've told your VCR to record two shows that overlap—or even "touch"—one of them will be incomplete or missing altogether. See page 64, step **5**, to review.

---

*Also a favorite mistake of VCR authors and other purported experts.

9. HOW TO COPY A TAPE FROM ANOTHER VCR OR A CAMCORDER

This chapter is to help you if you have a camcorder and want to make edited ("dubbed") copies of home videos for family or friends (or to blackmail siblings on yet-to-be-specified future occasions).

It often helps to work with another person, so that one of you can run the machine with the original tape (either it's a VCR or a camcorder) and the other can run the VCR recording the copy. Happily, you can hook together any combination of machine formats—VHS, Beta, 8-millimeter (more on these items shortly).

Dubbing well takes practice and patience. After a while, though, you'll be amazed at how much tighter and more interesting your edited camcorder tapes will be than the originals.

A word about the "other" kind of tape copying: It is wrong—and illegal—to copy rented movies. It's stealing.

Fortunately, many commercial rental releases are now produced with some sort of copyguarding to inhibit such illegal duplications (a system called Macrovision is the most commonly used). But even if there were no such safeguards, put yourself in the place of a movie maker: Imagine that you've spent *your* life's time and savings to pour *your* heart and soul into a movie—only to have it stolen via home-recording of "harmless" bootleg copies. End of sermon.

Hooking two VCRs together for tape duplication is really very similar to hooking up one VCR with a TV and/or cable: It's still just a matter of getting your **OUTPUT**s and your **INPUT**s right. Here's the idea:

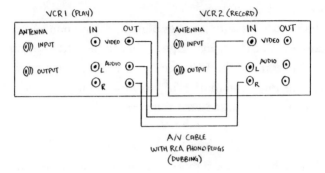

A/V CABLE
WITH RCA PHONO PLUGS
(DUBBING)

The machine playing the original tape must have its audio and video **OUTPUT**s connected to the audio and video **INPUT**s of the machine recording the duplicate tape.

To copy ("dub") from a camcorder (and it doesn't matter which format—8-millimeter, VHS, Beta, or VHS-C), it's essentially the same idea:

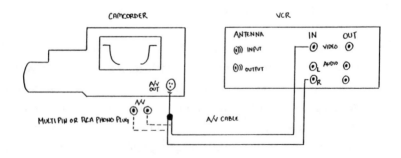

Then—with either of the above setups, you press **PLAY** on the machine with the original tape and **RECORD** on the machine making the copy.

Here's a complete checklist of the copying procedure:

1. Connect the outputs and inputs of your machines—video to video and audio to audio—as shown in the diagrams opposite (note: some camcorders have special multi-pin plugs that combine the separate audio/video cables shown here, but the idea is the same).

2. Put the tape to be copied—rewound or cued-up to the spot desired—into the *play* machine. If that machine has an **EDIT** switch, put it in the **ON** position.

3. Put a blank tape into the *record* VCR.

4. On most inexpensive VCRs, especially of recent vintage, plugging dubbing cables into the inputs automatically overrides the signals from the machine's tuner. However, older or more expensive VCRs have an **INPUT SELECT** switch (also known as **TUNER/LINE** or **TUNER/AUXILIARY**). This selection switch tells your VCR to ignore signals from its tuner and instead record whatever is coming through its inputs—handy, since it eliminates connecting and disconnecting the dubbing cables. So, if necessary, use the **INPUT SELECT** function of your *record* VCR to select the "line" or "auxiliary" input. Check your manual.

5. Set the recording speed of the *record* VCR—preferably to **SP** (or **Beta II** if you own a Beta machine). Don't go the cheap route here; it's penny-wise and pound-foolish to copy a treasured family videotape at a slower, lower fidelity speed.

6. Set up your TV to monitor the dubbing process the same way you'd normally watch a tape playing in the recording VCR.

7. Push **PLAY** on the *play* machine and **RECORD** on the *record* VCR.

8. If there's a part of the original tape that you don't want on the copy, simply push **PAUSE** on the *record* machine when that part appears; let up on the **PAUSE** button again when the unwanted part finishes. Of course, while your *record* VCR is in **PAUSE**, you can also fast-forward or rewind the original tape on the *play* machine, to cue up the proper spot.

9. If there's a part of the original tape that you want to copy in slow-motion or even freeze-frame, you can do this: Simply put your *play* machine into freeze-frame or slow motion, and the *record* machine will record this as it "sees" it—in freeze-frame or slow-motion.

A dubbed videotape will *never* look quite as good as the original, but there are a few things you can do to maximize the quality of the copies:

- Dub at the highest, fastest speed.

- Buy good, gold-plated dubbing cables. They're priced under twenty dollars and they really do decrease the loss of signal from the *play* machine to the *record* VCR.

- Use better grade videotapes. In twenty years, if the better tapes look even 5% better, that little bit of extra money will have bought you clearer memories.

- ***Video enhancers***, which decrease the loss of signal, are not a bad investment if you plan to do a lot of copying. They're available through many mail-order companies. Just be sure you know who you're dealing with, and that you can get a refund if the enhancer doesn't enhance.

10. SPECIAL EFFECTS
AND HOW TO USE THEM

One of the joys of watching any tape replay on your VCR is that you don't have to watch it in real time. You can stop the picture, move it ahead, back it up, slow it down—you're in complete control. The key to this magic is a basic knowledge of your VCR's special effects.*

- Rewind Search: If a group of people are watching a movie in a theater, and someone misses a line of dialogue, the Explanation-Noise Domino (END) Effect can be severe:

Alice:	What'd the good guy say?
Brian:	"Shhh!"
Carla:	"He said '………………' "
Dave:	"What did the bad guy say?"
Emily:	"I couldn't hear—Carla was talking."
Frank:	"He said '…………………' "
Gloria:	"Is this scene a flashback?"
Hal:	"I couldn't hear—Brian was shushing everyone."
Brian:	"Shhh!"

And so on.

But if you're all watching a tape at home, of course, you can use rewind search to back it up, and hear the line again: *While the tape is playing* (that is, do *not* press **STOP** first), push the **REWIND** button and hold it down until you get to the part of the tape in question. Then press **PLAY** again. (On most newer machines, you don't even need to hold down the **REWIND** button; the tape remains in rewind search until you push **PLAY** again).

*Note: Some of the newer VHS VCRs that don't offer the LP recording speed will *play* LP tapes OK, but if you try to use any special effects, the screen will go blank.

- <u>Fast forward search</u> works the same way—in the other direction. You'll find it indispensable for skipping through commercials, movie credits, etc. And you can see most of the action in a baseball game in about five minutes.*

  <u>Note</u>: Most VCRs have these visual search features, but a two-head VCR will show lines in the picture when you search.

- <u>Double-speed play</u> (or 2× play or fast-play) moves the tape at twice the normal speed—slower than fast-forward search, but good for speeding accurately through short bursts of tape to specific moments. It's also great fun with home videos—watching slow, matronly aunts play croquet like Keystone Kops.

- **PAUSE** (or "<u>freeze-frame</u>"). Use this when you want to freeze a video picture and look at it—a close play at first base, credits rolling at the end of a movie. This, too, is a lot of fun with a home video, because—as you know—*anyone* (even Dad) can be made to look extremely goofy on film if you catch the right moment.** But don't mis-use **PAUSE**. *Don't* use it to stop a tape while you answer the phone, make a sandwich, add on a room, etc. **PAUSE** is designed to show you one tape frame for a few seconds. When paused, the tape is stopped against the rapidly-spinning heads, which wears both the tape and the VCR. For long intervals, use **STOP**.***

*"Nah—just kidding... ...no—really! Wait! Hey! Put that bat down!......"

**Driver's license and passport photos, family reunion poses, wedding portraits, etc.

***Yes, **STOP** often takes an extra few seconds to get going again, forcing you to rewind a bit to avoid missing anything, but it's *a lot* easier on your equipment. Besides, many newer machines have reduced that start-up waiting period to just a fraction of a second, with a new feature called full-loading (covered later in this book).

- **FRAME-BY-FRAME** (or **FRAME ADVANCE**) is just a series of freeze frames—handiest for sports events or for analyzing a special effect in a movie.*

- **SLOW MOTION** is also just an extension of **PAUSE**, really. But instead of stopping entirely, the videotape simply moves more slowly across the head drum.

    <u>Note</u>: Both **PAUSE** and **SLOW MOTION** are considerably better on a four-head VCR than on a two-head model. With a two-head VHS VCR, you shouldn't expect much in the way of a usable picture in those special effects modes—at best a decent pause or slow-motion on tapes that have been recorded at the EP/SLP speed. Don't buy a two-head VCR if good-quality slow-motion is important to you.

---

*How *did* Linda Blair spin her head around in *The Exorcist*? Was she double-jointed, or what?

11. THE CARE AND FEEDING
OF YOUR VCR

Happily, the electronics in VCRs just keep getting better and better. Improvements in integrated circuit miniaturization have enabled the manufacturers to design one IC to do the work of three.

Mechanically, too, today's VCRs are much improved. Mass production has resulted in increased quality and reliability at lower cost. Overall, the $200 VCR on the shelf at your favorite store is a better machine than the $1200 monster from ten years ago.

Wonderful devices that they are, though, VCRs are still machines, which eventually break—and then usually cost you money to repair.*

But if you take care of your VCR, it will break down less, last longer, and give you a better overall return on your investment. It's just like a car in that respect: Today's cars are better built, too, but you still can't expect to drive one for 20,000 miles without ever lifting the hood. As your dentist has so often tried to drill into your head:

> *Proper preventative maintenance is the key*
> *to increased longevity and reliability.*

You:	"Can you fix it?"
Jeff, the Service Technician:	"Sure—no problem."
You:	"How much will it cost me?"
Marlon, the Other Service Technician:	"mmm...Let's see...... two-and-a-half hours' labor... one rhebob defibrillator...one wingdam demodulator... two half-caret VLSI chips... and a side of fries... —eleventy trillion dollars and 85 cents."
You:	*...(something's caught in your throat*
	—your tongue, maybe)...
Marlon:	"Plus tax."
Jeff:	"Will that be cash or check?"

Jeff and Marlon, technicians with a combined 931 years of experience in consumer electronics repair, are asked: "What typically goes wrong with VCRs?"   Their answer:

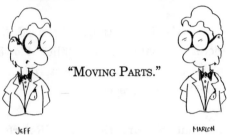

"Moving Parts."

Jeff

Marlon

Among the various moving parts, the *idler*, which physically drives the two reels of the videotape during rewind, fast-forward and play, naturally gets a lot of use and is the most common VCR problem.

Then there are mechanical timing adjustments, which involve the proper *alignment* of the gears, levers and guideposts of a VCR's *tape transport*—another oft cited repair.

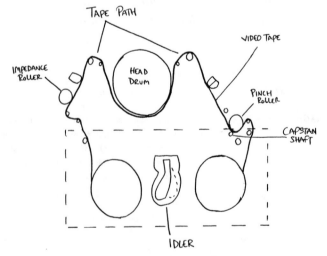

TAPE PATH

VIDEO TAPE

IMPEDANCE ROLLER

HEAD DRUM

PINCH ROLLER

CAPSTAN SHAFT

IDLER

"How to Keep Your VCR Away from Jeff and Marlon"

by Jeff and Marlon

1. Get it professionally cleaned annually—twice a year if it's used heavily or in the home of a smoker (smoke is a major cause of damage to VCRs—and lungs and clothes).

    A professional cleaning is more than just a perfunctory head cleaning. The entire tape path, guideposts and belts must also be thoroughly scrubbed of oxide buildup and other contaminants, and the entire chassis should be vacuumed free of dust.

    A VCR cleaning should cost you between $24 and $36; more than that and you're being overcharged. Less and the shop is probably performing just the head and tape path cleaning—running a simple head-cleaning cassette through the machine—which you can do.

2. If you use a head cleaner cassette, *follow all of its directions.* Neither Marlon nor Jeff recommends head cleaner cassettes—but not because they don't do an adequate job. Rather, it's because (surprise!) people don't follow the directions.

    Most head-cleaning cassettes are very abrasive to your VCR heads and should never be in the machine for over one minute.

    Jeff tells of one woman who didn't understand why her VCR heads needed $200 in repairs. "But I used a head cleaner!" she protested, "In fact, I had it in there for *over an hour!*"

    There are two types of head cleaning cassettes: wet and dry. The wet kind uses an alcohol-based solution—not as abrasive but not as thorough. The dry kind *must* be somewhat abrasive to work correctly, and abrading your video heads is just not a good idea. However, if used correctly (sparingly), it will not unduly harm your machine.

Marlon suggests this alternative to head-cleaning cassettes:

- Put a new videocassette in your machine;
- Fast-forward it to the end;
- Push **PLAY**;
- Put the tape into rewind search (as covered on page 75).

Since videotape is what video heads are built to withstand, a videotape moving opposite to the normal direction of the tape is a kinder, gentler way to dislodge most of the crud built up on the heads. But neither this nor a head-cleaning cassette eliminates the need for an annual professional cleaning.

3. Keep your VCR away from contaminants (smoke, dust, fumes from paint and solvents, etc.), heat and humidity.

4. Buy name-brand videotapes—with the stylized VHS logo. If it lacks the logo, it's probably a poor-quality knock-off.

5. Use your VCR at least once in a while. Many VCR owners believe that the less they use their VCR, the longer it will last—much like a car. That's *partly* true. But certain parts, like the rubber belts, need to be moved fairly frequently to keep them pliable (if you leave your car parked for a year, with all of the car's weight constantly on one part of each tire, well... it's not pretty). Use your VCR at least once a week, to keep things loose and pliable.

6. Use common sense: Keep irresponsible kids and pets away from it (the tape slot is for tapes only—not toy soldiers, building blocks, puzzle pieces, ice cream, doggie treats, etc.). Follow directions—and don't ignore warning symptoms.

7. Avoid cut-rate rental tapes. In the first place, rental movies are just about the worst thing for your VCR. As Marlon puts it, "You don't know where they've been." Maybe the previous renter was someone every bit as meticulous as you ("ah, it'll dry—I'll just wipe off the label"). Maybe not.

   Furthermore, a 99¢ rental price isn't as big a bargain as it seems. If two video stores pay $30 each for a copy of the same movie, but store A charges 99¢, while store B charges $1.99, store A has to rent that movie about 30 times more than store B before seeing any profit on it. Under the wacky assumption that each store wants to make a profit, this means you have 30 more chances at store A to become the hapless schmuck who rents "Pretty Woman" right after Mr. Sneaky Soda Spiller returns it.   Enough said.

8. Avoid excessive and unnecessary use of freeze-frame and slow motion. *Don't* confuse **PAUSE** with **STOP** (see page 76)!

9. Never force a videotape into your VCR. If it doesn't want to take it, there's a good reason.

10. If a tape gets stuck inside the VCR, don't try to remove it yourself. Take it to technicians who know what they're doing (no—not even Josh or Corey).

THIRD-DEGREE CRATTIBIP

12. WHEN IT'S TIME FOR A NEW VCR

While reading this book, maybe you've caught yourself saying, "Gee, that sounds like a nice feature to have." Or: "The new VCRs sound so much easier—maybe I should get one and save on headache remedies."

Maybe so.
- Is your VCR less than three years old?
- Does it have four video heads?
- Does it offer HiFi stereo?
- Can you record with it while you're there *and* from the timer?
- Is its performance and ease of use satisfactory to you?

If you answered "no" to three or more of these questions, you'd probably get your money's worth out of a new VCR—even if you're not a card-carrying couch potato. By owning a VCR you're more comfortable recording with, you can build up a nice stockpile of movies, miniseries, concerts, ballgames, or reruns of Gilligan's Island. In short, you'll have more high-quality programming ready when *you* have time to watch.

It's a nice, powerful feeling—that freedom of choice. But it all starts with knowing how to get your VCR to behave and to obey you. And if you can't do this with your current machine, maybe it's time to put it up in the kids' room and invest in a new one.

If so, here are a few guidelines on how to go about it:

Features You Really Shouldn't Be Without

Chances are, you'll own your VCR for five to ten years. The following features—available on even low-priced VCRs—will make those years much more pleasant. Don't buy a new machine now unless it has all of these features (machines with all of these start at $200-250):

- <u>On-screen programming</u> (see pages 62-67). Get the digital entry kind, and try it in the store to make sure you understand it.

- <u>Quartz tuning</u> (see page 41). Some cheapo machines still lack quartz tuning, but the few bucks you save just aren't worth it.

- <u>Index search</u>. This is the video equivalent of music search on a cassette deck. Whenever the VCR begins to record, it places an invisible marker on the tape, thus marking each separate recording for easy locating. So, if you have no idea what's on a particular tape—but it's indexed—the VCR will quickly look through the entire tape (in either direction) and find the beginning of every program. It's very easy to use and a terrific time-saver.

- <u>A real-time counter</u> usually goes hand-in-hand with the index search. The real-time counter keeps track of *tape used* and *tape remaining*—in hours, minutes, and seconds, instead of the older VCR method of counting revolutions of the tape. This means no more guessing how much time is left on a tape. And you can use the real-time counter to find your way through tapes that are not now indexed—simply by knowing the length(s) of the program(s) on the tape.

Features You Should Strongly Consider

These features are "essential" once you get them and see how terrific they are, but they're not actually essential if you're a casual user who mostly rents movies.

- <u>Four video heads</u>. As you now know, a four-head VCR gives you freeze-frame and slow motion, but it also records a better picture —roughly 20% sharper. This advantage is more pronounced when you record at the slower (lower quality) EP/SLP speed. Fast-forward and rewind search are clearer, too—some four-head VCRs barely show any "noise" bars in visual search.

- <u>HiFi stereo sound</u> doesn't do you any good without a stereo TV or stereo music system. Historically, TV sound, has been—to put it tactfully—lousy. And regular old VCR sound is even worse, especially tapes recorded at EP. By contrast, the clarity of a HiFi stereo VCR, played through *any* stereo system—no matter how old and decrepit—is stunning. So before you decide against a HiFi stereo VCR, go take a test-listen.

  One note here: A lot of folks—good, kind, decent folks whose kids grew up with rock-and-roll anyway—tend to associate "stereo" with words such as "noise," "loud" and "aspirin." <u>Not necessarily.</u> Stereo simply means better, more lifelike sound (stereo simply splits the sound *spatially* between left and right channels). The volume level is up to *you*. Stereo systems get "loud" when the Rolling Stones demand Satisfaction; but they're downright tasty when Harry James, Duke Ellington, or ol' Blue Eyes swing for you at normal, human decibel levels.

- Quick Access (or Quick Play, or Full Loading). The Smokescreen Institute* is having a field day with this one. The only way you'll know if a VCR has this feature is to try it.

On a VCR with quick access, once a tape has initially been put into **PLAY**, it's been loaded and wrapped around the video heads. This takes about three seconds. Thereafter, anytime you press **STOP** (*not* **PAUSE**; remember page 83!), then **PLAY** again, the picture returns to the screen in a fraction of a second (machines without quick access take the full 3 seconds to get re-wrapped).

To do the quick-access trick, the VCR keeps the head drum spinning with the tape positioned just a fraction of an inch away from the heads (if you wait more than about 5 minutes, the head drum stops this spinning and then it takes the normal 3-second time to wrap the tape around it and get up to speed).

This may not seem to be a very big deal, but if you're doing a lot of searching or dubbing, that three-second delay can be a rather large pain in the sitter.**

---

*The professional research and consulting firm hired by the VCR manufacturers to give the same simple feature five million different names.

**fanny

Features You Probably Don't Need

There are some merits in these VCR options, but they don't give enough value for the dollar to make them a good buy. You spend a big difference in money for smaller improvements and more specialized (and questionable) features.

- <u>Super VHS</u> (S-VHS). This may change, but S-VHS seems to be caught in a technological squeeze.

  Advantages: – VCRs equipped with S-VHS record a considerably better picture. The clarity on S-VHS tapes is outstanding,

  Disadvantages: – *S-VHS is too good for broadcast TV.* The TV/cable signals you receive have a picture with 330 lines of resolution (see the Glossary); S-VHS pictures have 430 lines, but you won't see this improvement when recording broadcast signals with 330 lines.

                                   – There aren't many S-VHS rental movies.

                                   – S-VHS machines are expensive—about $150-200 more for the S-VHS circuitry.

If you have a 30 inch or larger picture tube TV, or a projection TV, you will definitely notice the extra quality of S-VHS, and if you spent the money on a TV that expensive, then you can readily afford and justify the expense of the better format. But you don't need it—wait for recordable LaserDisc instead (that's like a CD player with pictures—even better than S-VHS).

- "Digital" anything (except digital tracking, which is a useful feature that adjusts your heads so that your tape tracks perfectly and therefore looks better).

People who own "digital" VCRs will disagree, mainly because they want to justify all the money they spent on such limited toys. Most digital VCRs let you either digitally freeze a live TV show or a frame of a videotape—or both. And—oh yes—you can also do things like create multiple consecutive images of a videotape (for studying your golf swing), but *really*.

The height of uselessness is the digital VCR that lets you watch a videotape *and* a live show <u>*at the same time!*</u>

Wow.

Um... why, exactly, would you want to do this? Isn't the whole purpose of a VCR so you can watch what you want *when* you want? If you've got one program on tape anyway, why would you want to watch it at the same time as a live broadcast?

A Shopping List

Now that you've got an idea of what's good and what isn't, go shopping:

- Check the video or consumer magazine of your choice and see what they recommend. Then use this as a *guideline*—and stay flexible. Often the specific models they recommend are already discontinued.

- Go into the stores and try a few machines with the features you want.

- Do some advance price checking and comparing. Don't go into a store expecting everything you want for $200. It's not going to happen. You get what you pay for.

- When you're in the stores, listen to the salesperson if he/she seems knowledgeable—but avoid the Digitenglish lecturer. In some stores, you'll know more about the machines than the sales people; they know only Digitenglish. Remember: If you buy a VCR and then run into problems using it, you'll want a salesperson who is willing and *able* to help—on the phone or in person.

- If prices are negotiable, try to get a better one (without bullying, which usually doesn't work).

- Be sure that the store you buy from has an adequate exchange/ refund policy, in case the VCR of your dreams turns out to be a nightmare.

- Be sure that the store you buy from will protect the price if they or one of their competitors puts your model VCR on sale shortly thereafter.

G. GLOSSARY

There are a lot of strange terms and acronyms in your VCR manual. Understanding them is important if you want to get the most out of your machine. This glossary explains the more common terms.

Antenna INPUTs: The places on the back of a VCR or TV where you can connect the antenna or cable. The VCR doesn't record "from" the TV; it has its own TV tuner built in that can tune in channels directly from the signal stream of the aforementioned antenna/cable. Some (older) VCRs and TVs have separate **VHF** and **UHF INPUT**s; some (newer ones) combine them.

Antenna OUTPUTs: The place on the back of your VCR from which it transmits signals to your TV (on the VCRs *output channel*: 3 or 4). Some older VCRs have separate **VHF** and **UHF OUTPUT**s; some newer ones combine them.

Audio dubbing: More common on camcorders than VCRs, this feature lets you replace the original soundtrack of your videotape with a new one. For example, you might want to replace the soundtrack of your favorite home video with sound effects, music or narration.

Audio-video INPUTs: When copying a tape, this is the place on the back of your VCR to connect the other VCR's (or camcorder's) audio/video **OUTPUT**s.

Audio-video OUTPUTs: Connect these to another VCR's audio/video inputs for tape copying; *or* connect them to a TV monitor for better picture and sound when viewing.

Automatic rewind: A feature—fairly standard now—that automatically rewinds a tape if you let it play all the way to the end.

Beta: Unfortunately for Beta-format VCR owners, Beta is essentially dead. In the early years of VCRs, it was a competing (and actually superior) alternative to VHS, but better VHS marketing led to Beta's demise—from all manufacturers except Sony. To verify that a machine is indeed a VHS format, look for the stylized "VHS" logo.

Beta I, II, III: The three recording speeds for Beta format VCRs. Beta I (found only in very old machines) allows 90 minutes of recording time on a standard L-750 cassette. Beta II gives you three hours; Beta III, 4.5 hours. As with VHS, the less recording time, the better the picture.

Bilingual: Also known as **SAP**, for Separate Audio Program, this is a separate channel feature offered by some stereo TVs and VCRs (see also: **MTS**). It lets the broadcasting station include another soundtrack, the idea being to broadcast in, say, Spanish in appropriate areas. So if you had a VCR with the **Stereo / Bilingual** switch you could select the **Bilingual** option and listen to a TV show in Spanish.*

Cable compatible: This means simply that the TV or VCR has the coaxial connector used for cable—the cable will screw right in.

Cable ready: This means that the built-in tuner can receive cable frequencies—in short, a VCR with at least 105 channels.

*A common mistake people make is accidently having their Stereo/SAP switch on their TV or VCR set to SAP. Unless your local station is broadcasting on the SAP channel, you might not notice that you're not getting the stereo, either. So be sure to check your machine if you have that switch and make sure it's in the Stereo position.

Cassette erasure tab: The little tab on the front edge of each VHS video cassette. If it's intact, you can record on the cassette; if not, you can't (rental movies always have the tab removed to prevent accidental erasure—and you should do the same for those priceless home movies). However, to record without the original tab, you can "mimic" it with a piece of masking or cellophane tape covering the opening.

Channel 3/4 switch: Usually located on the back of your VCR, this sets your VCR's ***Output channel***. To prevent interference from a local station broadcasting on 3 or 4, set the switch to 4 or 3, respectively.

Condensation: When a VCR is brought from a cold place to a warm one, condensation ("dew") may build up on the video heads. Most VCRs have a ***Dew light*** to warn you of this condition. DO NOT use your VCR if the dew light is on (most won't let you anyway). The videotape may stick to a wet head drum, ruining the tape and possibly damaging the VCR. So don't use your VCR if it's ice cold to the touch. Let it warm up to room temperature first.

Dew light: See ***Condensation***.

Digital: A term often overused ("New! Digital Cola!"), in VCRs, it refers to the use of computer circuitry to do one or more of the following: **(i)** enhance picture quality; **(ii)** allow the VCR to show two different picture sources at once (***Picture in Picture***); **(iii)** allow the VCR to store one frame of the videotape on screen while the tape continues to move—called "digital freeze." Most of these features are no big deal. So if a salesperson says a VCR is "digital," ask "Digital what?" If this causes a furrowed brow or stammering and blustering ("you know— *digital*, man! It's cool!"), ask to see the manual or another salesperson.

Digital tracking: A good, practical use of digital circuitry, this feature automatically aligns the tape and heads.

Digithead: Someone who knows a lot about electronics and likes to show off. Usually harmless, often obnoxious, always insomnia-curing.

Dolby stereo: This is the same Dolby as on your audio cassette deck, and it works the same way—masking out the inherent noise (hiss) on older, non-HiFi tapes (if you have a HiFi VCR, there's no hiss). Dolby stereo VCRs record soundtracks much as conventional VCRs do—in a straight line on the edge of the VHS tape—so the Dolby stereo machines are also called "Linear Stereo." Most non-stereo VCRs record their sound similarly, but on just one channel (monaural), whereas the "linear stereo" VCR divides the sound into left and right channels (stereo). But it's not great fidelity, hence the need for Dolby.

Double-speed play: This plays tapes at twice normal playback speed—good for zipping through dead spots in a ball game where **high speed search** might be too fast. It's also terrific for home movie fun.

EDIT switch: For better copies (dubs), put this switch *on* when copying from a VCR that has it to another one (whether it has it or not).

Eight-millimeter (8-mm): Another videotape format (as distinct from VHS and Beta)—not to be confused with the old 8-mm films. 8-mm video is primarily a camcorder format (yes, you can buy 8-mm VCRs but you won't find many rental movies). 8-mm is probably the best choice for a home video camcorder, because it uses a much smaller tape (and therefore machine), without losing the 2-hour recording time or good picture and sound. The tape is higher quality and ages better, too.

Electronic tuner: There are several types: If you're buying, look for a quartz tuner. There are some manually adjustable electronic tuners (on unusually low priced VCRs); avoid these like The Plague.

Express record: This is to a VCR what a sleep timer is to a clock radio: It lets you start recording manually but the machine shuts off after the amount of time that you designate—great for recording the end of the late movie without having to worry about the VCR being on all night.

Extended Play (EP): Also called ***SLP***, this is the recording speed on your VCR that gives you six hours of recording time on a standard T-120 cassette. There is a noticeable loss of picture quality, because the tape moves more slowly and you're squeezing more information on the same amount of tape, but on most of the newer machines with ***HQ***, the picture is more than adequate.

Flying erase head: Although this sounds like what Chuck Norris eventually does to the bad guys, it actually refers to an extra video head mounted on the head drum of a VCR or camcorder that erases one video frame at a time. This gives you considerably better and cleaner editing capabilities. Most the better camcorders have this; if you're buying a camcorder, strongly consider paying a little extra to get this feature.

Four heads: More expensive VCRs have four video heads instead of two. The benefit is better recording quality, especially at the EP/SLP recording speed, and also improved special effects such as slow motion, freeze frame, or visual search. Most people know the latter; very few the former, but a four head VCR does record a better picture.

Frame advance: When the VCR is **PAUSE'd**, you can advance the tape one video frame at a time—handy for second-guessing referees. For high quality pauses and frame advance, you need a four-head VCR.

Frame-by-frame: See *frame advance*.

Freeze-frame: Better known as **PAUSE**. Push the **PAUSE** button and the tape stops on that frame. On a two-head VCR, the picture won't be very clear, but most four-head VCRs have good freeze-frame—that's what the extra heads are for. Note: Using **PAUSE** for too long (more than a couple of minutes at once) can damage the tape because of the friction built up of the stationary tape against the spinning head drum.

Full-loading tape transport: A relatively new feature that reduces (by about 70-80%) the waiting time after you push the **PLAY** button.

GOTO: Moves the tape ahead a certain amount of time.

Heads: They do the recording and playback—four is better than two.

HiFi Stereo: All HiFi VCRs are stereo, but not all stereo VCRs are HiFi—look for the HiFi logo somewhere on the machine (a "Dolby Stereo" VCR does not sound nearly as good as a HiFi Stereo). VCRs with HiFi Stereo compare favorably to compact disc players in clarity and dynamics (louder louds, softer softs, and virtually no audible hiss —no need for Dolby). HiFi VCRs use an extra pair of heads that are mounted on the video drum and act like video heads, but they record only audio. If you have a stereo TV and/or a good stereo system that you can hook your VCR to, HiFi is a smart investment.

High-speed search: This feature (generally found on four-head machines) allows the VCR to speed up the tape as much as 200 times normal speed *while you watch it*. You use it primarily to skip through commercials or other unwanted sections of tape (most VCRs produce no sound during the process). Advertisers hate this feature.

Horizontal resolution: The number of *vertical* lines—as counted *horizontally*—in a television screen's picture. See also **Resolution**.

HQ (High Quality): Any one of four circuits—invented and licensed by JVC—that strikingly improve record and playback quality. To be considered HQ, a VCR must have at least one of these four HQ circuits, named in proper Digitenglish:

- A *Detail Enhancer* (*DE*) increases the amount and quality of fine details in a video recording;

- *Luminance Signal Noise Reduction* (*YNR*) reduces the noise, or jittering, along the edge of a video image;

- A *White Clip Circuit*, (*WCL*), produces a higher black and white frequency range, giving a sharper, more detailed picture;

- A *Chrominance Signal Noise Reduction* circuit improves color quality by reducing color noise and shimmer, which can be especially noticeable in large solid fields of one color.

Indexing/Index search: One of the best innovations of the last few years, this is the video equivalent of music search on an audio cassette deck. Indexing lets the VCR hop from recording to recording, by locating the invisible electronic mark the machine makes whenever you push the **RECORD** button.

Instant play: See ***Full loading***.

Instant Timer Record: See ***Express record***.

Intro search: An extension of the **INDEX** feature, it tells the VCR to search, in either direction, recording by recording, for the one you want, showing you the beginning of each recording.

Linear Stereo: See ***Dolby stereo***.

LP (Long Play): On VHS machines, it's the recording speed that allows four hours of recording time on a standard T-120 cassette.

Monitor: A TV with a special circuit that allows the video portion of your VCR or computer to play separately from the audio portion. A monitor generally offers improved resolution (clarity).

Multichannel Television Sound (MTS): This mouthful simply means "stereo TV"—sound produced in two channels for more lifelike spatial separation—like FM radio or your cassette deck, CD player, or your turntable. Watching a TV program in stereo is much more realistic, especially if you connect your TV or VCR (whichever one has the MTS) into your stereo system.*

Noise: Video noise is a disagreeable pattern of color flecks or graininess on a videotape, usually caused by dirty video heads, a poor quality tape, and/or poor antenna or cable reception. You'll also get more noise in the EP/SLP recording speed.

---

*It's just like being at the theater—except you can put real butter on your popcorn, and if some lardhead is telegraphing each scene before it happens, you can kick him out because it's your house.

One-Touch Record: See *Express Record*.

On-screen programming: A much easier—and increasingly afford-able—method of setting the VCR's programmable timer to record a TV show when you aren't going to be there. Next to quartz tuning, this is the next best reason to spend a few extra bucks on a VCR.

Output channel: The channels that VCRs and cable converter boxes use to send their tuned signals to a TV (on VCRs: channel 3 or 4; on converter boxes: channel 2, 3 or 4).

PAUSE: See *Freeze frame*.

Picture-In-Picture (PIP): Available on some digital VCRs, PIP allows you to watch a live televised program while also watching a tape in a corner of the screen, or vice-versa (but why?). This is not to be confused with the PIP offered on some digital TVs, where you can watch two *live* TV shows at the same time.

Quartz tuning: A type of electronic tuner, increasingly common in the latest models of VCRs, that automatically tunes in the stations the VCR is receiving. It's worth its weight in gold; most of the horror stories you hear about VCRs are really not about hooking up the machine but about trouble with tuning in the stations.

Quick Access: See *Full loading*.

Quick Timer Record (QTR): See *Express record*.

Real-time counter: This is another new and fairly wonderful feature. Until recently, VCRs simply counted revolutions of the tape reels to indicate where in the tape you were—not very exact and difficult for archival storage of programs. But a real-time counter allows the VCR to keep track of tape elapsed in true hours, minutes and seconds.

Resolution: The sharpness of a TV picture is determined by how many *vertical* lines—counted *horizontally*—form the complete image. The typical VCR image has about 240 vertical lines of resolution; a TV broadcast has 330 lines; S-VHS VCRs can reproduce 430. But an image is only as sharp as its original source: a S-VHS VCR can't record 430 lines from a 330-line broadcast. There is a huge competition among manufacturers to make products with high resolution numbers that don't offer much practical value but that look impressive to consumers.

SAP (Separate Audio Programming): See **Bilingual**.

Six heads: If a VCR really has six heads, it's a four-head VCR with the extra two heads for HiFi stereo recording capabilities (there were, once upon a time, VCRs with five *video* heads, but never six; the most they have now is four).

Simulcast switch: A switch on some stereo and HiFi stereo VCRs that allows you to record the picture from a TV station while simultaneously recording the sound from an FM radio station (concerts used to be simulcast quite a bit before the advent of stereo TV).

Slow motion: A feature chiefly found in four-head VCRs—good to have if you're a sports fan or a camcorder owner. Obviously, it's the ability to play a tape at a slow speed.

SLP: The slowest recording speed on a VHS VCR—allowing 6 hours on a T-120 tape. See ***Extended Play (EP)***.

SP: Standard Play—the fastest recording speed (2 hours on a T-120 tape)—the speed that gives the best picture.

Stereo TV: See ***Multichannel Television Sound (MTS)***.

STILL FRAME: See ***Freeze***.

Super VHS (S-VHS): A new type of VHS VCR with a much higher resolution (i.e. a terrific picture). Instead of 240 lines of visible resolution, S-VHS has 430. However, to see this, you need a higher-resolution screen *and* signal (not currently available on the air or from cable). Tapes recorded in S-VHS won't play in most regular VHS machines; an S-VHS machine will play regular VHS tapes just fine (but with no better resolution). So, as yet, S-VHS probably isn't practical unless you have at least a 30-inch screen and some extra disposable income of which you need to dispose.

Tape speed: The SP, LP, and SLP/EP recording speeds on VHS machines; the Beta I, II, and III recording speeds on Beta machines. When you record, you must set one of these speeds (the machine will match this speed automatically during playback).

Tracking: The control that adjusts the position of the tape on the heads. If the "tracking is off" you'll see lines in the picture and often (especially on HiFi machines) hear a garbled sound. Simple, correctable tracking problems are often mistaken for faulty tapes or VCRs.

TV/ Video switch (or ***TV/ VCR switch***): The switch on your VCR that selects between the respective pictures your VCR and TV are "seeing."

Two heads: A two-head VCR is your basic low-priced machine, fine for occasional recording and movie rentals, but not so great for better quality recordings or special effects (freeze frame, slow motion, etc.). For that you need four heads.

Two-times play (***2×play***): See ***Double-speed play***.

UHF INPUTs: See ***Antenna INPUTs***.

UHF OUTPUTs: See ***Antenna OUTPUTs***.

Varactor tuner: A nasty old type of electronic tuner—thankfully outdated—with a little set of little wheels and little switches. You put your little fingers on the little wheels and turn them each a little until you find each little station you want—a little tedious.

VCR: Video Cassette Recorder. You know—that Nice Warm Place For The Cat To Sleep.

VCR Plus+: A nifty little hand-held remote control made by Gemstar Development Corporation. It's an electronic brain that, when properly programmed, assumes all of the functions of your VCR timer. You just enter a five-to-nine-digit code on the VCR Plus+ keypad for the show you want to record, and it will turn on the VCR (and cable converter box, if necessary) at the right time, select the right channel, record the show, and then shut off. You don't even have to set your VCR's clock.

Vertical lock: Also called **V. Lock**, it's for the freeze-frame feature: If there's "jitter" in the frozen picture, then your VCR has been drinking too much coffee, and you adjust the video lock control until the still picture is steady. Most two-head VCRs don't have this control.

VHF INPUTs: See **Antenna INPUTs**.

VHF OUTPUTs: **Antenna OUTPUTs**.

VHS. This stands for Video Home System—the trademarked name coined by JVC and Sony when they patented the tape back in 1976. It's pretty much synonymous with "VCR" now, but that wasn't always the case. Look for the familiar "VHS" logo when you buy blank tapes; tapes without this symbol may be inferior or even damage your machine.

VHS-C: A compact form of VHS cassette, used primarily in camcorders. When inserted into the appropriate adapter, a VHS-C tape will play in a regular VHS machine.

Video/TV switch: See **TV/Video switch**.

Video dubbing: The process of replacing the *picture* (video) portion of a recorded tape with some other video—without disturbing the tape's audio portion. Some manufacturers don't make this distinction clear, or they refer to video dubbing as the erasure and replacement of *both* video and audio (which is really what most people want, anyway).

VISS (**Video Index Search System**): See **Indexing**.

Visual search: See **High-speed search**.

Index

About the Author

Dave Murray, who has been selling VCRs for seven years, felt
so sorry for his confused customers that he was inspired to write
a book about them (the VCRs, not his customers). Dave's humor
column, "Out of the Ordinary," ran for a year and a half in the
alternative weekly newspaper, the *Syracuse New Times*, and
generated much enthusiasm and fan mail, mostly from close
relatives. Dave lives in his hometown of Syracuse, New York.
This is his first book.

About the Illustrator

Joe Congel's passion for cartooning is superseded only by his
passion for his TV remote control. Joe's comic strip, "Tropical
Pete," appeared in the *Syracuse New Times*, and he also worked
for a time as advertising art director for a chain of video rental
stores. This is his first book, and he hopes it will set off a chain
reaction of job offers so plentiful that he'll be forced to put down
the remote control. Joe, his wife, Jackie, and daughter, Rita,
live in Syracuse.

By the way, if you liked this book, there are many others that you—or someone you know—will certainly enjoy also. Here are descriptions of a couple of them:

A Little DOS Will Do You

Here's a fast, easy way to get up-to-speed on DOS—the heart of your IBM or IBM-compatible personal computer! This short, little book covers **Version 4.01** and any earlier versions of DOS, giving you quick, hands-on lessons on: disks, directories, shells, files, pathnames, menus and more. As you progress through the lessons, you'll build and modify your own files and directories.

Concise and WordPerfect

Never before was so much information put so clearly into so little a package! This handy little book applies to both WordPerfect **versions 5.0 and 5.1**, offering practical examples on entering, formatting and saving text, using columns, blocks, tabs, footnotes, page numbers, outlines, macros, spell-checking, and more. There's also a handy reference guide in the back cover. Don't miss it!

See the next page for a full title list—the order forms are provided. Or, you can contact us for further information on the books and where you can buy them locally:

Grapevine Publications, Inc.
P.O. Box 2449
Corvallis, Oregon 97339-2449 U.S.A.
Phone: 1-800-338-4331 (Fax: 503-754-6508)

Item #	Book Title	Price
	Personal Computer Books	
29	A Little **DOS** Will Do You	$ 9
28	Lotus Be Brief	9
32	Concise and **WordPerfect**	9
30	An Easy Course in Using **DOS**	18
38	An Easy Course in Using **Lotus 1-2-3**	18
37	An Easy Course in Using **WordPerfect**	18
40	An Easy Course in Using **dBASE IV**	18
35	The Answers You Need on the **HP 95LX Palmtop PC**	9
34	**Lotus** in Minutes on the **HP 95LX Palmtop PC**	9
	Hewlett-Packard Calculator Books	
19	An Easy Course in Using the **HP 19BII**	$ 22
22	The **HP-19B Pocket Guide:** Just In Case	6
20	An Easy Course in Using the **HP-17B**	22
23	The **HP-17B Pocket Guide:** Just In Case	6
05	An Easy Course in Using the **HP-12C**	22
12	The **HP-12C Pocket Guide:** Just In Case	6
31	An Easy Course in Using the **HP 48**	22
33	**HP 48** Graphics	20
18	An Easy Course in Using the **HP-28S**	22
25	**HP-28S** Software Power Tools: **Electrical Circuits**	18
27	**HP-28S** Software Power Tools: **Utilities**	20
26	An Easy Course in Using the **HP-42S**	22
	Curriculum Books	
14	Problem-Solving Situations: A Teacher's Resource Book	$ 15
	Consumer Books	
36	House-Training Your VCR: A Help Manual for Humans	$ 8

(Prices are subject to change without notice)

Grapevine Publications, Inc.

626 N.W. 4th Street   P.O. Box 2449

Corvallis OR, 97339-2449

For orders and order information:

Phone: **1-800-338-4331** (503-754-0583) Fax: **503-754-6508**

To Order **Grapevine Publications** books:

☎ **Call** to charge the books to **VISA/MasterCard**, *or*

✍ **Send** this Order Form to: Grapevine Publications, P.O. Box 2449 Corvallis, OR 97339

Qty.	Item #	Book Description	Unit Cost	Total

Shipping Information:

❑ For orders <u>less</u> than $7 .... **ADD $ 1.00**
  or

❑ **Surface Post** shipping/handling .... **ADD $ 2.50**
  (allow 2-3 weeks for delivery).......................... *or*

❑ **Priority Post** ❑ **UPS** shipping/handling ....**ADD $ 4.00**
  (allow 7-10 days for delivery) .... *or*

❑ **International Air Mail:**
  Add $5 <u>per book</u> to Canada and Mexico. Add $10 <u>per book</u> to all
  other countries (allow 2-3 weeks for delivery).

Subtotal	
Shipping See shipping Info.	
TOTAL	

Payment Information

❑ **Check** enclosed (Please **make your check** payable to **Grapevine Publications, Inc.**)
  (International Check or Money Order must be in U.S. funds and drawn on a U.S. bank)

❑ **VISA** or **MasterCard #**_____ Exp. date_____

Your Signature _____

Name_____ **Phone (**) _____

Shipping Address_____
  (Note: UPS will not deliver to a P.O. Box! Please give a street address for UPS delivery)

City_____**State**_____**Zip**_____**Country**_____

Reader Comments

We here at Grapevine like to hear feedback about our books. It helps us produce books tailored to your needs. If you have any specific comments or advice for our authors after reading this book, we'd appreciate hearing from you!

Which of our books do you have?

Comments, Advice and Suggestions:

May we use your comments as testimonials?

Your Name: Profession:

City, State:

How long have you had a VCR?

Please send Grapevine Catalogues to these persons:

Name _____

Address _____

City _____ State _____ Zip _____

Name _____

Address _____

City _____ State _____ Zip_____